Education, Gender and De

This compelling book takes a novel approach to the complexities of girls' and women's education in the global South. To unravel the critical issues and processes behind educational advancement and to identify the factors that support the construction of educational well-being and agency from a gender perspective, the book narrates the stories of women who have successfully built their educational careers to higher education level. The book creatively applies the human development and capabilities approach to analyse and assess educational advancement and development.

Mari-Anne Okkolin offers a fresh voice to the field of education, gender and development. The book draws on rich, in-depth evidence from Tanzanian women who have reached higher education, placing them amongst a very small percentage of women in the Tanzanian and sub-Saharan contexts. The book explores the women's school experiences, everyday life practices and familial arrangements, and the values, expectations and assumptions associated with education and the schooling of girls and women.

Due to the multi-disciplinary nature of the book, it will be of great interest to multiple academic audiences: postgraduates, researchers and academics. It is of particular relevance for all those interested in education, sociology, development studies, gender/women's studies and qualitative research methodology. The book will appeal especially to scholars working with the capabilities approach. It will also be of value beyond academia, for education practitioners in planning and implementing education and equality policies internationally.

Mari-Anne Okkolin is Postdoctoral Fellow at the Centre for Research on Higher Education and Development, University of the Free State, South Africa, and Researcher at the Department of Education, University of Jyväskylä, Finland.

Routledge International Studies of Women and Place

Series Editors: Janet Henshall Momsen and Janice Monk, University of California, Davis and University of Arizona, USA

For a full list of titles in this series, please visit www.routledge.com/series/SE0406

Education, Gender and Development

A Capabilities Perspective

Mari-Anne Okkolin

NEW YORK AND LONDON

First published 2017
by Routledge
711 Third Avenue, New York, NY 10017

and by Routledge
2 Park Square, Milton Park, Abingdon, Oxon OX14 4RN

First issued in paperback 2018

Routledge is an imprint of the Taylor & Francis Group, an informa business

Library of Congress Cataloging in Publication Data
A catalog record for this book has been requested

ISBN 13: 978-1-138-62419-1 (pbk)
ISBN 13: 978-1-138-67304-5 (hbk)

Typeset in Times New Roman
by Taylor & Francis Books

To Aino and Lauri – because stories matter

Contents

Illustrations

Acknowledgements

When I obtained my PhD in 2013, I thought that the project on which this book draws was then completed, as projects are by definition. However, my dissertation developed further into this book, which meant some more time spent with the study. My first encounters with 'case Tanzania' took place in 2002 and ever since we have spent occasional, concentrated and highly intensive time together. Our children were born at the time of this process; hence, for them 'mom's Tanzania things' have always been there. Evidently, this has been a long project and process, and my gratitude is due to many people.

First of all, without the ten special, courageous, confident and open Tanzanian women who shared their lives with me, this book would have not existed. I sincerely hope that I have done justice to your stories and achievements. Now, having a few years' retrospective perspective on my doctoral studies, I've come to value even more the wisdom and patience of 'my' Professor Jyrki Jyrkämä and Tapio Aittola. I've also come to appreciate even more the passion and enthusiasm of Dr Elina Lehtomäki, supervisor, colleague and friend, for global education and development issues. I feel privileged to have had the opportunity to collaborate with some of the leading experts in this area of research, and I want to thank Professors Rauni Räsänen, Elaine Unterhalter and Melanie Walker. I wish to give my special recognition to Elaine and Melanie (the founder and leader of the Centre for Research on Higher Education and Development (CRHED) at the University of the Free State, South Africa) for their valuable input and encouraging comments about my work and scholarly thinking. Besides, this book would not exist without Melanie's suggestion and support. I have had an amazing two years as a post-doctoral fellow at the CRHED and I have so much enjoyed and benefitted from our discussions and debates. As observed by a CRHED colleague in her book, the intellectual environment at the CRHED is just amazing. I couldn't agree more and I want to thank you all for creating and being part of it.

At different phases of the research process I have been supported by so many institutions and people: the study has been funded by the Academy of Finland, the Nordic Africa Institute, the Emil Aaltonen Foundation, and the Department of Education at the University of Jyväskylä; Nawa and Peter, thank you for your time and work; many thanks to my brilliant colleagues and friends at the Department of Education and Department of Social Sciences and Philosophy; and my very special thanks to our Tanzanian partners, especially the School of Education at the University of Dar es Salaam, and Professor Eustella P. Bhalalusesa. And Magreth and Hanna, it has been my absolute pleasure to have this joint (ad)venture with you!

As tends to be the case, my special thanks are due to those closest to me. I hope that you all know that I know that without you I would not have made it. The biggest appreciation belongs to Timo, my husband. Thank you for being my 'critical counterpart' and beloved companion for years – gosh, decades :) ! And finally, Aino and Lauri, thank you for bringing all the joy and happiness into my life. Our son, who was nine months old when he visited Tanzania for the first time said the other day: 'I've been thinking that when I grow up, I don't think I'll travel that much anymore'. Seriously, my dearest ones, it is all up to you! Yet, I do hope that you and your sister have learned to understand why stories matter, many stories matter.

Abbreviations

BEST	Basic Education Statistics in Tanzania
CA	Capability Approach
CEDAW	Convention on the Elimination of All Forms of Discrimination Against Women
COSTECH	The Tanzania Commission for Science and Technology
CSO	Civil Society Organisation
DFID	Department for International Development (UK)
ECOSOC	Economic and Social Council
EFA	Education for All
ESR	Education for Self-Reliance
ETP	Education and Training Policy
ESDP	Education Sector Development Programme
FAWE	Forum for African Women Educationalists
FEMSA	Female Education in Mathematics and Science in Africa
GAD	Gender and Development
GER	Gross Enrolment Ratio
GPI	Gender Parity Index
HE	Higher Education
IMF	International Monetary Fund
MDGs	Millennium Development Goals
MoEC	Ministry of Education and Culture
MoEVT	Ministry of Education and Vocational Training
NEPAD	The New Partnership for Africa's Development
NER	Net Enrolment Ratio
NSGRP	National Strategy for Growth and Reduction of Poverty
OECD	Organisation for Economic Cooperation and Development
PCB	Physics, Chemistry and Biology
PCM	Physics, Chemistry and Mathematics
PEDP	Primary Education Development Plan
PETS	Public Expenditure Tracking Study
PSLE	Primary School Leaving Examination
SDGs	Sustainable Development Goals
SEDP	Secondary Education Development Plan

SIDA	Swedish International Development Cooperation Agency
SSA	Sub-Saharan Africa
SWAP	Sector-Wide Approach
TPR	Teacher-to-Pupil Ratio
TSH	Tanzanian Shilling
TTC	Teacher Training College
UDHR	Universal Declaration of Human Rights
UDSM	University of Dar es Salaam
UN	United Nations
UNICEF	United Nations Children's Fund
UNDP	United Nations Development Programme
UNESCO	United Nations Educational, Scientific and Cultural Organization
UPE	Universal Primary Education
URT	United Republic of Tanzania
WB	World Bank
WFP	World Food Programme
WHO	World Health Organization
WID	Women in Development

Part I
Setting the Scene

1 Introduction

> The single story creates stereotypes, and the problem with stereotypes is not
> that they are untrue, but that they are incomplete. They make one story become
> the only story.
>
> (Chimamanda Ngozi Adichie, TEDGlobal 2009)

Leyla, one of the ten highly educated Tanzanian women whose stories are
told in this book, described her everyday life experiences as follows: 'my
brothers talked about school, and other issues, what is happening in the town
and world issues, while I was with my mother in the kitchen preparing the
meal'. Leyla's story reflects the idea of the education and schooling of girls
and women in her family. At the same time, it resonates with the broader
socially constructed *idea of Tanzanian woman*, including the idea of adequate
and appropriate education for girls and women.

   In many developing countries, the socio-cultural expectations for the future role
of girls, that is, marriage and family, the high social status attached to marriage
and motherhood, and the conflicting gender ideologies at the community
level, diminish the demand for female education and promote gendered differences in educational opportunities and outcomes. This is evident, in particular,
when moving from the lower to the advanced and higher levels of education.
As reported in the final analysis of the Millennium Development Goals era,
the largest disparities in enrolment ratios are found in tertiary education and
the most extreme disparities are those at the expense of women in sub-Saharan
Africa (UN 2015a). This challenge and shortfall is explicitly acknowledged in
Goal 4 of the 2030 Agenda for Sustainable Development, adopted by the
United Nations General Assembly on 25th September 2015 (UN 2015b)[1].

   Apart from the evident challenge regarding the widening gender gap and
declining trend in girls' and women's educational progression, female students'
educational outcomes in the global South are characterised by non-enrolments,
overage enrolments, absenteeism, poor motivation, low self-esteem, poor academic performance, high levels of dropout, high levels of illiteracy and, consequently, limited labour-market opportunities. This kind of story is heard often
from Tanzania, the country-case selected for this book to represent and exemplify
the problem of education and gender globally. But, alongside the hegemonic

development discourse and education policy narratives, different stories exist. In this book, the stories of Leyla, Genefa, Hanifa, Amana, Amisa, Rabia, Wema, Tumaini, Rehema and Naomi are told. At the heart of the book lie their experiences and insights into critical issues and processes behind educational success.

In her famous TED talk entitled 'The danger of a single story', Chimamanda Ngozi Adichie emphasises: 'of course, Africa is a continent full of catastrophes [...] but there are other stories that are not about catastrophe, and it is very important, it is just as important, to talk about them.' Hence, 'stories matter – many stories matter' (Adichie 2009). In 2006 when I initially met and spoke with the women whose stories are represented in the book, only 2 per cent of Tanzanians were enrolled in higher education and only 30 per cent of Master's degree students were women. Evidently, only a tiny proportion of Tanzanian women had succeeded in reaching the university level of education, including the women who participated in my research. This makes their educational career and stories, by definition, a contestation of the *idea of Tanzanian woman* and a process of 'becoming something else'. Indisputably, women's stories are packed with problems and barriers, but despite all the challenges, against the odds, they were able to reach higher education. How? This is what the book is about and the reason why their stories matter.

Fukuda-Parr (2012) has argued for the importance of understanding the power of narratives and discourse in international development arenas. She claims that the overarching Millennium Development Goals (MDG) narrative, for example, is fundamentally about the normalisation and institutionalisation of particular development ideas and bringing those to the forefront of policy dialogue. Without going into the international development debate and criticism of the MDGs here, it is of relevance to recognise the concerns that are directed at the quantification and oversimplification of complex development challenges (see also Unterhalter 2014). Subrahmanian (2005) has made a statement that echoes Fukuda-Parr's remarks, according to which, achieving 'substantive equality' in education depends essentially on two processes to reveal *what* has been reached and realised, and *how*. Therefore, she proposes, in the analysis and assessment of gender equality in education, the focus should be on the socially constructed pathway(s) to equality, referring particularly to the quality of experience.

What the arguments made by Fukuda-Parr, Unterhalter and Subrahmanian suggest is that, in the planning, implementation and assessment of global policies, that is, institutionalised mainstream development narratives, too little attention is given to the notions of social justice and equality of opportunity and treatment, let alone to the understanding of dynamics and underlying power relations in any human endeavour, including education. What they assert suggests further that it is of critical importance to learn from those who are 'the targets' of global and national goals. As an example, Nobel laureate economist Amartya Sen (2005, xiii) has commented that the denial of 'voice' from the common people is just elitist, cynical and tends to encourage impassivity; for him, 'participation in arguments is a general opportunity, not a particular specialised skill (like composing sonnets or performing trapeze arts)'. With respect to this critique, within the research on education, gender and development, as in the research on development in general, the research strand that gives emphasis to people's 'voices' has intensified its status.

One of Nussbaum's (2001) central points in addressing the 'upheavals of thought' is that the stories we tell ourselves about who we are and what we feel shape our emotional and ethical reality. Likewise, reflecting the purpose of storytelling and the social responsibility of the writer, Susan Sontag (2007) emphasises how a 'writer of fiction both *creates* [...] a new world, a world that is unique, individual; and *responds* to a world, the world the writer shares with other people but is unknown or mis-known by still more people, confined in their worlds: call that history, society, what you will'. Sontag proposes: 'the writers who matter most to us are those who enlarge our consciences and our sympathies and our knowledge'.

All the critical remarks presented above consider essentially the question of whose voice and story are of relevance to be told and heard. In doing so, they call for broadening the horizon from one story to the richness of stories and acknowledging the value and power of narratives. In this book I wish to respond to these demands. Drawing on rich, in-depth evidence, the book narrates the stories of ten highly educated Tanzanian women who have constructed their educational careers all the way to higher education. At the core of the book are the questions: How did they do it? How did they manage to achieve various educational 'beings and doings', in contrast to the majority of Tanzanian women; and, what factors supported and enabled them to pursue and realise their educational aspirations? This directs the analytical interest towards the women's school experiences, everyday life practices and familial arrangements, and to the values and attitudes attached to the education and schooling of girls and women – reflecting the expectations and assumptions attached to the *idea of Tanzanian woman*. The book illustrates also the extent to which the women had the freedom to exercise educational agency; that is, to have *de facto* alternatives, autonomy to make decisions and to take actions accordingly. My argument is rooted in the capabilities approach (CA) initiated by Amartya Sen and philosopher Martha C. Nussbaum (see e.g. Sen 1985, 1993, 1999; Nussbaum 2000; see also Comin and Nussbaum 2014). The approach asserts that *well-being* and *agency* are equally important and interdependent aspects of human life; yet they are ontologically and analytically distinct.

This book is interdisciplinary in nature and contributes to the social and human sciences in various ways. By giving 'voice' to the women, my aim is to take 'into consideration not only the global benchmarks, but also, and most importantly, the situation on the ground' (Lehtomäki et al 2014). The book thereby complements the understanding of gender equality and equity in education, gained on the basis of the hegemonic narrative, which guides the way that governments collect and present statistical data and international organisations make comparisons. Different from most of the writings on gender, education and development, which are characteristically problem-oriented, my aim is also to provide positive portrayals of women in sub-Saharan Africa. In addition, I wish to demonstrate how the conceptual frame of the human development and capabilities approach can be operationalised and empirically applied, towards which much of the capabilities criticism has been directed. Therefore, the outcome of the book is capabilities-informed analysis and understanding of the complexities and intersecting issues of girls' and women's education and schooling in the global South, with a special focus on women's narratives and female agency.

The book is contextualised into Tanzania and the country provides a macro-case for the research. Yet the insights and evidence discussed in the book are of critical importance to be taken into consideration in the planning and implementation of education policies in other parts of sub-Saharan Africa. Apart from the relevance to the global South, due to the adaptation of the capabilities approach in particular, the book contributes to the debates on educational advancement and assessment internationally.

An essential outcome of the book is its contribution to the qualitative research literature. Among others, Mauthner and Doucet (2002, 2003) argue how the research methods that we apply are not neutral techniques but are based on theoretical, epistemological and ontological assumptions. Hence, the research encounters, as well as our methodological choices, are sites where some voices may be enhanced and others silenced. In this book, women's stories represent such a qualitative and interpretative research paradigm, which places emphasis on people's life-worlds and 'voices' (see Okin 2003 and Nussbaum 2004). At the same time, their narratives advance such epistemological and methodological engagement in the educational debate, which validates the voices and subjectivity of those who are the 'targets' of equality and equity policies (see Unterhalter 2007; Walker 2010). Evidently, by choosing *what* to study and *how*, the researcher can present the particular story that she wants to be told and heard. To phrase it differently, deciding what aspects of the phenomena to pay attention to impacts on the selection of 'indicators', the data collection and analysis methods, and, in consequence, on the data *per se*, the results, and, finally, the scientific claims that can be drawn on these grounds. Because of the implications of scientific premises and positionings, and to advance transparency of the qualitative research process, various methodological and ethical considerations, subjectivity and evaluative aspects alike that have been taken into account alongside the study will be explicitly represented.

I have organised this book into four parts. Part I consists of chapters that elucidate the study on which the book is based: it introduces briefly the phenomena of education, gender and development *par excellence*, the ten highly educated Tanzanian women and the research process, including the researcher. In this section I thereby argue *why* to address education, gender and development, *how* this has been empirically carried out, and *by whom*. As a research monograph and the first study that explicitly combines the use of the capability approach with sociologically informed engagement with reflexive hermeneutic premises, I devote especially large space to the deliberation of scientific presuppositions, research methodology and conceptual considerations.

The chapters in Part II cover the background needed to understand the field of study. I start with a broader discussion and research on education, gender and development (Chapter 5) and continue by locating the gendered educational challenges and achievements in the Tanzanian context (Chapter 6) The primary emphasis in Chapter 7 is given to an explanation of principal theoretical and conceptual approaches to address and assess gender equality in education and to a positioning of my study in the frame of 'equality of capabilities'. Thereby, this chapter also involves an explanation of how the capabilities approach is analytically applied in the study.

The largest part of the book consists of three empirical chapters in which the women's experiences and insights are represented in detail. At the very core of the book is Part III, which accounts for the women's school experiences (Chapter 8), social and familial experiences (Chapter 9) and agency notions (Chapter 10). Within the school environment, the themes of school infrastructure, human relations and the learning environment are revealed. In Chapter 9, the analytical lenses are angled towards women's everyday life and customary practices. The purpose of this chapter is also to find out about women's *ideas* of themselves, 'Us', and what enabled them to overcome educational barriers and reach higher education, what made the difference for the majority of others, 'They', around them.

Chapter 10 is about women's agency, aspirations and freedoms, which are examined on the basis of the question: Who decides? This chapter aims to illustrate the extent to which the research participants had the freedom to exercise educational agency; in what ways their 'beings and doings' were in accordance with the socially constructed *idea of Tanzanian woman*, including the hegemonic idea of adequate and appropriate education for girls and women; and how these women aimed to become something else. Therefore, the focus of analysis is on the processes through which educational choices are made, as contextualised and pre-conditioned by socio-cultural structures (discussed in the two previous chapters). The concluding chapters in Part IV briefly summarise the key findings and discuss the meaning of 'story telling' methodology for educational policy planning, implementation and assessment for 'transforming our world for sustainable development' (UN 2015b).

Note

1 The selected education and gender targets and indicators of the Millennium Development Goals and Sustainable Development Goals relevant to this book are listed in the Appendices. Also, Education for All (EFA) Goals and the full preamble of the Incheon Declaration are included in the Appendices.

Bibliography

Adichie, C.N. 2009. The Danger of a Single Story. TEDGlobal Talk. July 2009. Interactive transcript https://www.ted.com/talks/chimamanda_adichie_the_danger_of_a_single_story.

Comin, F., and M.C. Nussbaum, eds. 2014. *Capabilities, Gender, Equality. Towards Fundamental Entitlements.* Cambridge: Cambridge University Press.

Fukuda-Parr, S. 2012. *Recapturing the Narrative of International Development.* Geneva: UNRISID.

Lehtomäki, E., Janhonen-Abruquah, H., Tuomi, M. T., Okkolin, M.-A., Posti-Ahokas, H., and Palojoki, P. (2014). Research to engage voices on the ground in educational development. *International Journal of Educational Development,* 35, 37–43.

Mauthner, N. and Doucet, A. 2002. Knowing Responsibly: Linking Ethics, Research Practice and Epistemology. In Mauthner, M., Birch, M., Jessop, J. and Miller, T. *Ethics in Qualitative Research.* London: Sage, 123–145.

Mauthner, N. and Doucet, A. 2003. Reflexive Accounts and Accounts of Reflexivity in Qualitative Data Analysis. *Sociology* 37, no. 3: 412–431.

Nussbaum, M. 2000. *Women and Human Development: The capabilities approach.* Cambridge: Cambridge University Press.

Nussbaum, M.C. 2001. *Upheavals of Thought: the intelligence of emotions.* Cambridge: Cambridge University Press.

Nussbaum, M.C. 2004. On Hearing Women's Voice: A Reply to Susan Okin. *Philosophy & Public Affairs* 32, no. 2: 193–205.

Okin, S.M. 2003. Poverty, Well-Being, and Gender: What Counts, Who's Heard. *Philosophy & Public Affairs* 31, no. 3: 280–316.

Sen, A. 1985. Well-being, Agency and Freedom: The Dewey Lectures 1984. *Journal of Philosophy* 82, no. 4: 169–221.

Sen, A. 1993. Capability and Well-being. In *The Quality of Life,* edited by M. Nussbaum and A. Sen, 30–53. Oxford: Oxford University Press.

Sen, A. 1999. *Development as Freedom.* Oxford: Oxford University Press.

Sen, A. 2005. *The Argumentative Indian: Writings on Indian Culture, History and Identity.* London: Penguin.

Sontag, S. 2007. *At the Same Time: essays and speeches.* New York: Farrar, Strauss and Giroux.

Subrahmanian, R. (2005). Gender equality in education: Definitions and measurements. *International Journal of Educational Development,* 25, 395–407.

UN. 2015a. *The Millennium Development Goals Report.* New York: United Nations.

UN. 2015b. *Transforming our world: the 2010 Agenda for Sustainable Development. Resolution adopted by the General Assembly* on 25 September 2015. New York: United Nations.

Unterhalter, E. 2007. *Gender, Schooling and Global Social Justice.* London: Routledge.

Unterhalter, E. 2014. Walking Backwards into the Future: a comparative perspective on education and a post-2015 framework. *Compare,* 44, no. 6: 852–873.

Walker, M. 2010. Capabilities and Social Justice in Education. In *Education, Welfare and the Capabilities Approach,* edited by H.-U. Otto and H. Ziegler, 155–179. Opladen: Barbara Budrich Publishers.

Wilson-Strydom, M., and M.-A. Okkolin 2016. Enabling environments for equity, access and quality learning post-2015: Lessons from South Africa and Tanzania. *International Journal of Educational Development,* 49: 225–233.

2 Why and How to Address Education, Gender and Development

Education has made me who I am but my family background made me know the importance of education.

(Amisa, 26)

As previously pointed out, in the global South, the educational opportunities and outcomes of girls and women are generally lower than those of boys and men. This is the case in particular when moving from the lower to the advanced and higher levels of education. In 2006, when I initially met with the women whose stories are narrated in this book, 49 per cent of Tanzanian students enrolled in primary education were female; 47.5 per cent of first-year secondary school students and 40.5 per cent of the advanced level of secondary education students were women; 30 per cent of the Master's degree students were female. Evidently, there is a declining pattern in female students' access to education, retention rates and attainment in the education system. With only 1.4 per cent of Tanzanian people enrolled in higher learning institutions in 2013, it is clear that only 'one in a million' Tanzanian women have reached higher education (HE). As noted in the 2013 national Basic Education Statistics, still the majority of students enrolled in HE are males (GER 7.9 per cent in comparison to 3.5 per cent for females), and still, the total enrolment is extremely low compared to the theoretical school age population of 20–23 years old[1] (URT 2013).

To put the Tanzanian women's achievements into perspective in the sub-Saharan African context, a brief statistical glimpse is helpful. A cross-national higher education report in Africa (Bunting, Cloete and Van Schalkwyk 2014), which included flagship universities in eight sub-Saharan African countries (including Botswana, South Africa, Tanzania, Mozambique, Ghana, Mauritius, Uganda and Kenya), pointed out that during their 11-year follow-up period from 2001 to 2011, the total (female and male) number of Master's student enrolments grew rapidly in most of the universities. However, in Tanzania, at the University of Dar es Salaam (UDSM), selected for the study because of its status as the 'most prominent public university in its country' (ibid.), the total dropped by 76 per cent over the five-year period, from 2,165 (N) in 2007 to 522 (N) in 2011. According to the report, in 2011 only three universities had 50 per cent or more female students in undergraduate programmes, while four universities, including UDSM, had undergraduate female proportions below 40 per cent; the average female enrolment in doctoral programmes across the eight universities was 37 per cent. At UDSM, the proportion of female students enrolled in the undergraduate, Master's and doctoral programmes were 39 per cent, 37 per cent and 28 per cent respectively.

Globally, the widening gender gap and declining trend in female students' educational progression is influenced by in-school and out-of-school factors. In addition to the factors related to the school environment, girls' and women's poor educational outcomes have been explained by political and institutional factors (supply) and by socio-cultural and socio-economic factors (demand). The influence of social factors on the demand for female education is clear when we look more closely at the parental and familial decision-making on whether to invest in female education or not; similarly, gender ideas and ideologies at the household and community levels may promote differentiated educational opportunities and outcomes for females and males (Odaga and Heneveld 1995; Unterhalter et al. 2014). In Tanzania, socio-economic and socio-cultural factors have been a part of policy discussions for over 40 years; they were first raised soon after the country gained independence in 1961 (Buchert 1994, Mbilinyi 1991); however, recent studies emphasise that similar challenges still exist (e.g. Colclough, Al-Samarrai and Tembon 2003; Okkolin, Lehtomäki and Bhalalusesa 2010; Unterhalter and Heslop 2011).

Therefore, in the analysis of education and schooling, *and* to tackle gender inequalities in education, it is of critical importance to pay attention to agencies, institutions and social relations at various levels of the education establishment (Unterhalter 2003a; Colclough, Al-Samarrai and Tembon 2003; King, Palmer and Hayman 2005). Depending on positioning, different levels and aspects may be emphasised: one option is to look at the education and training system itself; another possibility is to refer to the wider non-educational environment outside the educational establishment. Analysis of the non-educational environment may expand further to the legal, economic and political levels and to social and familial arrangements (Figure 2.1).

The environments that enable girls' and women's education and schooling, and factors that impact on female students' educational advancement, well-being and agency, are intertwined in various and complex ways. Clearly, attending school is not only about schooling as such; quite the contrary: female education and schooling is impacted on by political and institutional factors and even more importantly by socio-economic and socio-cultural factors. The policy initiatives which aim to reduce the gender gap and inequalities in education are mainly implemented in-school and targeted towards the school-related factors, in line with the mandate and scope of educational policy. However, on the basis of knowledge and understanding of the complexities and intersecting factors that have an impact on girls' and women's schooling and educational advancement, this emphasis alone is insufficient and needs to be complemented with the factors that are essentially

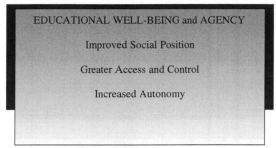

Figure 2.1 Process Towards Educational Well-being and Agency (modified from UNDP 2001).

part of the social and familial levels of the enabling environment – as seen in Amisa's remark above.

Apart from the challenge that arises from the singular focus on in-school environments to advance female educational advancement, the difficulty is based inherently on the very understanding of gender equality and equity in education. Research on education, gender and development has shown how different *understandings* of gender, equality and education generate different *approaches* with which to pursue gender equality in education, and hence different ways in which to *assess* development and change. The mainstream understanding, approach and assessment of social development – including gender equality and equity in education – is dominated by policy-informed, macro-oriented and quantitative-based approaches. However, as suggested by the concepts of 'gross and net enrolment rates' and the 'gender parity index', for instance, they only measure the equality in terms of amounts and numbers. Yet gender equality and equity only rests on, but is neither synonymous with nor an index of, gender parity. Consequently, if educational development and advancement is to be comprehended more as educational well-being[2] and agency notions are brought into the agenda (Figure 2.1) – as is the case in the international and Tanzanian policy narratives and discourses – this approach alone is inadequate and needs to be complemented.

Reflecting international policy commitments, the current educational policies in Tanzania encompass both gender parity and gender equality understandings, with quantitative and qualitative objectives (URT 1995, 2008). In consequence, both kinds of approaches and assessment are needed to realise equality and equity in education. The challenge embedded in the understanding of 'substantive equality' in education lies, however, in disclosing the complex relationships between structure and agency, that is, the process of 'what has been reached and realised and how', as previously discussed. On the one hand, this implies the need to understand the enabling and/or constraining social structures which constitute both the context and the outcome of people's well-being and agency; on the other hand, it necessitates understanding the critical counterpart for the structure, that is, the subjective self: presumably, an intentional and rational agent, aiming at something and capable of making rational judgements and choices. This book is based on such a methodological relationalism, which, by definition, draws attention to positions constituted by social relations. Hence, the women's educational well-being and agency (Figure 2.1) is understood to be socially constructed, as an interplay between social conditioning and agential responses; as a process, in which the women posit themselves and make educational choices in relation to their socio-cultural structures, that is here, their school environments and social and familial environments, which all reflect *the idea of Tanzanian woman.*

In my study I have been essentially interested in the 'structures of opportunity'. Women's narratives, in turn, represent the multifaceted negotiations and processes through which individual human beings locate themselves in social structures. Different from the macro-oriented and quantitative-based approaches, and to complement the information and knowledge base with which to assess educational advancement, a qualitative and actor-centered research methodology has been employed. To capture subjective insights into critical issues and processes behind educational advancement, and to identify the factors that support the construction of educational well-being and agency, the stories of highly educated Tanzanian women have been collected. Hence, this book is substantially about the experiences and insights of ten Tanzanian women, and their narratives are at the heart of the book. In the next section, I introduce the women and the study on which this book draws.

Notes

1 Population in 2006: 38.2 million; July 2015: 51 million (est.), CIA/World Fact Book: https://www.cia.gov/library/publications/the-world-factbook/geos/tz.html
2 For instance, the 2010 Human Development Report reaffirms the concept and the approach of human development and re-examines the distribution of well-being for inequality, gender equity and poverty (UNDP 2010).

Bibliography

Buchert, L. 1994. *Education in the Development of Tanzania 1919–1990.* Athens, OH: Ohio University Press.

Bunting, I., N. Cloete, and F. Van Schalkwyk. 2014. *An Empirical Overview of Eight Flagship Universities in Africa: 2001–2011. A report of the Higher Education Research and Advocacy Network in Africa (HERANA)*. Cape Town: Centre for Higher Education Transformation.

Colclough, C., S. Al-Samarrai, P. Rose, and M. Tembon. 2003. *Achieving Schooling for All in Africa: Costs, Commitment and Gender*. Ashbourne: Ashgate.

King, K., R. Palmer, and R. Hayman. 2005. Bridging Research and Policy on Education, Training and Their Enabling Environments. Special issue, *Journal of International Development* 17, no. 6: 803–817.

Mbilinyi, M., and P. Mbughuni, eds. 1991. *Education in Tanzania with a Gender Perspective: Summary Report*. Dar es Salaam and Stockholm: SIDA.

Odaga, A. and W. Heneveld. 1995. Girls and School in Sub-Saharan Africa, From Analysis to Action. World Bank Technical Paper No. 298. Washington D.C.: World Bank.

Okkolin, M.-A., E. Lehtomäki, and E. Bhalalusesa. 2010. Successes and Challenges in the Education Sector Development in Tanzania: Focus on Gender and Inclusive Education. *Gender and Education* 22, no. 1: 63–71.

UNDP. 2001. *Learning & Information Pack: Gender Analysis*. UNDP: Gender in Development Programme.

UNDP. 2010. *Human Development Report. The Real Wealth of Nations: Pathways to Human Development*.

Unterhalter, E. 2003a. Crossing Disciplinary Boundaries: the Potential of Sen's Capability Approach for Sociologists of Education. *British Journal of Sociology of Education* 24, no. 5: 665–669.

Unterhalter, E. and J. Heslop, 2011. *Transforming Education for Girls in Nigeria and Tanzania. A Cross-country Analysis of Baseline Research*. London, UK: ActionAid.

Unterhalter, E., A. North, M. Arnot, C. Lloyd, L. Moletsane, E. Murphy-Graham, J. Parkes, and M. Saito. 2014. *Intervention to enhance girls' education and gender equality. Education Rigorous Literature Review*. London, UK: DFID.

URT. 1995. *Education and Training Policy (ETP)*. Dar es Salaam, Tanzania: The Ministry of Education and Culture.

URT. 2008. *Education Sector Development Programme (ESDP) 2008–2017*. Dar es Salaam, Tanzania: Ministry of Education and Culture.

URT. 2013. *Basic Education Statistics in Tanzania (BEST) 2009–2013 National Data*. Dar es Salaam, Tanzania: Ministry of Education and Culture.

3 Introducing the Ten Highly Educated Women

In 2006, when we met for the first time, seven of the ten women were finishing their Master's degree studies. Most of the women were in their early thirties: two of them were aged 32 and two 33. The youngest woman was 26 and the oldest was 53; the other participants were aged 28, 30, 36 and 40 (some biographical characteristics are summarised in Table A.1 in the Appendices). Leyla, Genefa, Hanifa and Amana[1] were the first research participants[2] whom I met and I introduce them here accordingly.

Leyla is the oldest of this group of women, aged 53[3] at the time that we met. She belongs to the age group in Tanzania that used to carry out one-year national (military) service, which she undertook after finishing her advanced level of secondary education. Leyla has three children and she is a single parent, as she is widowed, although she had been divorced from her husband prior to his death. She is a Kiswahili and Geography teacher by profession and had worked for a long time in a boys' school. Her mother was a farmer and she had four years of primary education; further schooling was not considered to be necessary. Leyla's father was employed by the government and he used to work as a field assistant (training other farmers). He had six years of primary schooling and some agricultural studies as his educational background. Leyla (the second child) has four sisters and two brothers; in addition, her cousin used to live with them. Most of her siblings completed secondary school, and the firstborn supported the education and schooling of his younger brother and sisters remarkably.

Genefa, aged 40 when we met, is the second oldest within this group of women. She is married with four children. Her husband is employed and posted elsewhere; consequently Genefa has the main responsibility for their household and for bringing up the children. Genefa also used to work as a Kiswahili teacher in a secondary school. Her mother was a housewife and she had not attended school, only some adult education courses. Genefa noted that she could barely read or write. Her father was a police officer, but after an accident, he was no longer able to work. Hence, Genefa and her siblings, her two sisters and two brothers, were mainly supported by their oldest sister (also a police officer).

Hanifa (33) is married and she already had one daughter in 2006. She is an English teacher by profession. Hanifa is the fourth child in the family and has four sisters and two brothers. In addition, her uncle and their grandmother used to live with them. Her mother was a housewife and had three years of primary education schooling; her father had been to the teacher training college and had some further education. He used to work as a head district education officer. All of Hanifa's siblings are educated, but none of them to university level.

Amana (36) has two children and they are living with their father and his new family. Amana is also a teacher by profession. She has four sisters and two brothers, and the extended family members living with them included one cousin. Amana's mother had four years of primary education (which Amana said was typical at that time), was a housewife and took care of the family shamba (field). Her father was a veterinary surgeon. He used to work in the town, and came back to the village at the weekends. All of Amana's four sisters are teachers and her oldest brother has a Master's degree.

Amisa is the youngest of the research participants, aged 26 when we met. She co-habits with her boyfriend. Amisa is the firstborn in her family and has two sisters and two brothers: the last-born was enrolled in Form IV in secondary education and three others were doing their Master's degrees. Her mother and father got divorced and all of the children were living with the mother, who took care of them without any support from the father. Her mother holds a Bachelor's degree and works as an accountant. Apart from the nuclear family, Amisa's aunt and cousins used to live with them. They were remarkably assisted by her grandparents; namely, the parents of Amisa's mother.

Rabia (30) and Amisa are friends from university and I carried out a paired interview with them. At the time we met, she had a boyfriend. Rabia had worked as a teacher, as had many of the research participants. Her mother had gone to teacher training college and worked as a primary school teacher; her father had gone to university and had a degree in biology and chemistry; he worked as a secondary school teacher ; and two of Rabia's four sisters also worked as teachers. Rabia comes from a rather remote village and they used to have a big farm that was mainly kept for the family's needs. They had a large extended family, and apart from her grandparents, some other relatives also used to live with them.

The parents of Wema (32) are divorced, and, like Amisa, Wema too, together with her younger sister and older brother (who passed away in an accident), had lived with and were raised and nurtured by their mother, with no assistance from their father. Wema's family was supported by their uncle. Apart from the two siblings, two of their cousins lived with them. Wema's mother holds a Master's degree and she was employed in the teaching profession; similarly, Wema's father had a Master's degree, and he too worked as a teacher. Her late brother was a doctor and her sister has a university degree. At the time we met, Wema was married.

Tumaini (33) was introduced to me by Wema. She had worked as a programme manager in the areas of nutrition and health care and HIV/Aids, and when we met in 2008, she was finishing her Master's Degree in Public Health (to complement her previous B.Sc. Degree in Home Economics and Human Nutrition). She has three sisters, although the oldest one has passed away; apart from her sisters, three of her cousins lived with their family. Her mother has a certificate in teaching and she teaches political science at primary school. Tumaini's father is a prison officer; thus, she used to live in the prison camp staff quarters. "Maybe Std VII", that is, seven years in primary education, is what Tumaini considered that her father had obtained in terms of formal schooling. One of Tumaini's sisters is following in her footsteps and she has just finished her first degree; the other did not perform so well, according to Tumaini, and her parents are paying for her to go to college for a certificate in accounting.

Rehema (28) was living with her father and stepmother when we met in 2008. She holds a Bachelor's degree and did not have plans to continue further, at least at that time; she worked as a teacher and a university lecturer. Rehema is the firstborn in the family and she has one sister and two brothers. Both of her biological parents are teachers by profession; her step-mother also used to work in a teacher training college but when we met, she was a housewife and had a small clothing business; her father also made a living from his businesses. Apart from the immediate family members, one cousin and one nephew were living with them. Rehema had had an accident as a child, resulting in a physical impairment, but this was not severe enough to prohibit her from functioning.

Naomi (32) is married and in 2008, she already had a son. Naomi's background is also in education and she used to work as an English teacher. She is the firstborn in the family and has three younger sisters. Their aunt used to live with them. Naomi described how they used to live in 'educational surroundings', because of the educational and professional backgrounds of their parents: her mother holds a BA and her father has a PhD; her mother (retired) used to work as an executive educational officer, and her father (retired) as a priest, but also because her parents used to run an education centre. One of her sisters has a doctorate, and two others were studying at the university at the time of our conversation.

To add and summarise the research participants' familial backgrounds, a few more remarks are useful: firstly, the women diverged in their backgrounds with regard to geographical location and ethnic group; secondly, they represent different socio-economic groups, yet most of them described their childhood families as middle-class, having had a normal and moderate standard of living; the educational background of their mothers varied from no schooling at all (poor writing and reading skills) to Master's degrees, and their fathers' backgrounds from six years in primary education to doctorate level; they came from families with three to six children, and basically all of the siblings were educated to secondary education level, and in most cases beyond that; in the case of three of the childhood families, their parents had divorced and two of the women were basically raised and taken care of by their mothers; one had lived with and been supported by her father, with whom (and with his new wife) she was also living at the time we met; at the time of our conversations, three of the women were living with their boyfriends and one was dating but living alone in her own house. The other five were married with between one and four children, although two of them had divorced: one (later widowed) had dependent children, while the other participant's children were living with their father's new family.

At the time we met, all of the research participants were doing their Master's degree studies, barring one, who held a Bachelor's degree from the university; five of them were finishing their theses in education, two in development studies, and one in mass media and communication.[4] In addition, two women held B.Sc. Degrees in Home Economics and Human Nutrition, although one was just completing her M.Sc. Degree in Public Health. Eight of the women were qualified teachers. Apart from Wema, whom I encountered and conversed with in Finland, I met the women in Dar es Salaam in 2006 and 2008. In the next chapter I explain how the research process was empirically carried out.

Notes

1 In representing the women's stories I am using pseudonyms to protect their anonymity and identities.
2 I deliberately use the term research participant, rather than interviewee, let alone 'an informant', echoing Reinharz's (1992) assertion regarding feminist researchers, who are 'stretching the boundaries of what constitutes research'; as 'some [of us] choose to use personal voice, while some see the research as self-reflective, collaborative, and attuned to a process, oriented to social change and designed to be for women rather than only of women' (Reinharz 1992, 268–269). I strongly favour the idea of 'for women' and endorse Reinharz's suggestion that we may 'learn from the women that were studied, not only learn about' (ibid. 264).
3 The ages of the women at the time of the interview are indicated here and also in the annexed Table A.1. Later, after every citation, the name and the age of the woman are provided, but not, however, in the body of the text.
4 All of them have now completed their degrees.

Bibliography

Reinharz, S. 1992. *Feminist Methods in Social Research.* Oxford: Oxford University Press.

4   Introducing the Empirical Study

The starting point of the study on which this book draws, the thematic interest in education, gender and development in general, and Tanzania in particular, was pragmatic and guided by development cooperation and development policy. The reason for this was that in 2003, when I was first introduced to this 'country case', I used to work for a consultancy company that specialised in the area of education and development, and we were tendering for the support services of the Ministry for Foreign Affairs of Finland for the Education Sector Development Programme (ESDP) in Tanzania. Evidently, my understanding of the phenomena was based on and instructed by a particular kind of literature, and indeed, the very reason for becoming more interested in this area of study was to question the information and knowledge base used to put development politics and policies into practice, and to assess development and change. In other words, I started to ask questions about the validity of the mainstream understanding of gender equality and equity in education as the basis of development policy formulation and hence the basis of implementation and assessment.

After the unfinished tendering process, we submitted a research proposal to the Academy of Finland, succeeded in getting funds for a two-year pre-study, and commenced the macro-level policy research focusing on the gendered achievements and challenges of education sector reform in the country. Later, in order to develop an understanding of education, gender and development further, the policy-informed pre-study was complemented with a study that was first and foremost concerned with women's educational experiences and insights.

Pre-Study – Policy Informed Understanding

To initiate the pre-study, which was guided by development cooperation and policy , and to obtain an understanding of the status of gender and education in Tanzania, the following questions were posed: to what extent and how is gender addressed in the education sector in strategic planning, implementation, monitoring and evaluation; what are the strategies, means and instruments to promote gender equity in the education system; what has been achieved and what remains a challenge from the gender perspective? To carry out a situational analysis and to identify information gaps concerning gender in the most recent education sector development processes, policy documentation provided by the government of Tanzania and international development agencies was collected and analysed.

The data for the study was acquired through three fieldwork periods in Dar es Salaam. The purpose of the first preparatory visit in 2005 was to meet different stakeholders (policy – research – practice), that is, education officials, researchers in the field of education, gender and development, Tanzanian and international civil society organisations (CSOs), UN agencies (UNICEF and UNESCO in particular) and other partners in education and development. In addition, to deliberate on the critical issues regarding education sector reform in the country, the purpose of the first fieldwork period was to collect educational data, relatively difficult to obtain otherwise; and to become acquainted with the recent Tanzanian research and studies on education and gender, even more difficult to identify and acquire from outside the country. The corpus of the policy data for the pre-study consisted of the most recent education sector reforms in Tanzania and was limited to include data from the year 2001 onwards, when the Primary Education Development Plan was introduced. In addition, the analysis included yearly publishing related to the global processes of the EFA and MDGs, to which the Government of Tanzania is committed.

It goes without saying that both macro/micro, and quantitative/qualitative approaches have their advantages and disadvantages; without claiming that educational decision-making is based on macro-level and statistical information and knowledge base alone. However, as previously discussed, it is important to acknowledge the practical implications of different gender equality and equity understandings: the choice of what to measure and assess is a political process and differs from one stakeholder to another; and deciding what aspects of change to pay attention to impacts on the indicators, data collection and analysis methods; which produces not only different kinds of data, but also different kinds of results. Consequently, as discussed above, by choosing *what* and *how* to study, the policy-makers, researchers and practitioners can present the particular story they want to be told and heard (Moser 2007; see also Mauthner and Doucet 1998, 2003; Mauthner, Birch, Jessop and Miller 2002; Unterhalter 2003b, 2007a). As pointed out by Moser (2007), 'hard figures' produced by quantitative methodologies are crucial in building the case for addressing gender differentials, even if, as she (ibid.) observes, these figures are often contested and subject to interpretation; whereas referring to Reinharz (1992), hard figures display the trends over time regarding, for example, gender gaps in education. Different from 'number-based understanding', qualitative methodology enables in-depth examination of social processes, social relations, power dynamics and the quality of gender equality and equity.

My study is bound considerably to the policy-level and 'numbers-informed' pre-study findings. Yet to identify factors that enable educational advancement, well-being and agency, and to develop an understanding of the complexities and intersectionalities of girls' and women's education and schooling, such an approach was, simply, incomplete and inaccurate and needed to be complemented with a study that draws on the qualitative 'actor- and voice-centered' research paradigm.

Collecting Women's Narratives

In order to develop an understanding of education, gender and development further, the policy-informed pre-study was complemented with the study, in which the focus was categorically and by definition on the women's experiences and insights, that is, on their educational narratives. This book draws essentially on this data. In deciding upon the variety of data-collection methods, I ended up choosing semi-structured thematic interviews, due to the fairly specific themes and issues I had in mind to be discussed. I developed my interview themes on the basis of the much-cited review by Odaga and Heneveld (1995), which will be presented and discussed in detail in Part II. I phrased my interview questions Building on their categorisation of political and institutional factors, school-related factors and socio-economic and socio-cultural factors that have an impact on gender inequities in education, and following their comprehensive and detailed list of barriers and problems for girls' and women's education and schooling in the developing countries. Factors grouped by Odaga and Heneveld (ibid.) were complemented with critical issues and strategic priorities drawn from the analysis of policy documentation (to be discussed also in Part II). However, the idea was to focus more on *opportunities*, rather than problems and barriers for female education. In fact, conversations would be a more exact term to describe our encounters. For instance, and to avoid leading and instructing our conversations excessively, I phrased my questions into forms such as: 'What do you think, what kind of discussions did your parents have regarding the education of their children?', and 'Would you please tell me about your first memories at school?'

During my second visit to Tanzania in 2006, I met and spoke with six research participants. The discussions we had were retrospective in nature, although some future orientation was included (see the interview themes in Appendices). Four of my research participants were university colleagues and knew one another; I had not met them before. My impression was, however, that all our encounters were informal, open and pleasant, to which reflection I return later. Before the third fieldwork period, which took place in 2008, a seventh highly educated Tanzanian woman was interviewed in Finland, first of all, to hear her narrative, but also to test the slightly revised structure of the interview. Later that year, in Tanzania, three more women were interviewed. In total I have spent approximately five months in the country; hence, I have only visited Tanzania and have not lived there longer than one month at a time. Lastly, I visited Tanzania in January 2013 and had the opportunity to meet with some of the women. With some of the other women, I have stayed in touch through occasional e-mails, and have had more regular encounters with some of them through social media. Thus, I do know how seven of the women have constructed their educational, professional and familial pathways up to date; however, none of these semi-formal and informal conversations are included in the data and analysis *per se*.

Each of the interviews lasted an average of two hours and all of them were conducted in English. English is neither my nor the women's mother tongue, but there are basically two reasons for choosing English over Kiswahili: first, I am not familiar with the language, and second, I assumed that English is not going to be an issue and a problem in having discussions with these women, who had received a great part of their education and work experience in English. The assumption turned out to be, however, not as straightforward (as will be noted later in the section focusing on the women's learning environments), although not too critical, let alone prohibitive. Along with the growing number of cross-cultural studies, there is a considerable amount of literature arguing and debating the advantages and disadvantages of using the research participants' mother tongue as opposed to a language that they all know and share; taking advantage of interpreters, not only as language interpreters but also as cultural interpreters; and having a translator working with interview transcripts (see e.g. Bujra 2006; Holstein and Gubrium 2003; Ryen 2003). In my study, the use of Kiswahili would have definitely been a more crucial concern to assess, if information was being acquired, for example, from younger women and girls, who would not have been studying and working in English for quite as many years (cf. Posti-Ahokas and Okkolin 2015). Interestingly, we all found ourselves, the women and me alike, in the same situation of not using our mother tongue. Pietilä (2010), for instance, has raised interesting aspects to be considered in conducting research in a foreign language and cultural setting different to one's own. On the one hand, as the interviewee always acts as an ambassador of her/his own culture, in the best cases she/he may describe, explain and reflect on issues in great detail which might have been disregarded and taken for granted if the interviewer came from the same background; in that sense, using a foreign language in interviews may act as a key to understanding cultural and social differences embedded in cross-cultural encounters. As an example, one of the women used quite a lot of expressions such as: you know; I need to share this with you; stay with me; you see, which may be interpreted as an endeavour to properly open up the topic under discussion. On the other hand, there is a risk of glossing over things (no problems anywhere at all), or hypercriticism, or defensiveness ('you cannot understand since you come from elsewhere and are not living here'); similarly, there is chance for open and free airing of one's ideas (ibid., 415–417). As has been noted by Packer (2011) conducting fieldwork is never merely a contact between an individual and a culture, let alone the researcher becoming a member of the culture, but always the meeting of two cultures and two forms of life. This directs us to the salient and perennial insider–outsider question, and naturally, from the language question perspective, I do not know what I might have heard and learned if all the tones and nuances of the mother tongue had been available and in use.

The four women that I met and interviewed initially were proposed to me by our research project's Tanzanian advisor. During our encounters, I met two of the women together: while I had a conversation with one, the other was writing her educational story. With these four women we also had a group discussion. The fifth of the participants I met in Finland, as an exchange student at the time of completing her Master's degree studies, and she introduced me to her friend in Tanzania. These two women were interviewed as a pair in Dar es Salaam. During my third visit to Tanzania, one of the research participants was introduced to me by a Finnish colleague, and the second one again by the Tanzanian research advisor. Another contact to get in touch while in Dar es Salaam was suggested by the Tanzanian woman I talked to in Finland. As described, all of the research participants have been gathered through personal contacts. The only criterion I had was that the women were either university degree students or had completed their studies recently, and no additional requirements regarding their age, marital status, geographic origin, religion etc. were imposed. As a matter of fact, I found out interestingly and by chance that two of the women were Muslims (all of the eight others were Christians). I learned this accidentally, because when we met, it was the time of Ramadan, and naturally they politely abstained from any water or soda I served – not to mention any Marianne-candies I had as an ice-breaker. After gathering all of the narratives, the video tapes were transcribed for me by a female research assistant; then, the 'listening' to the women's educational experiences and insights began.

Analysing Women's Narratives – Voice-Centered Relational Method

In order to analyse the collected interview data, I applied the voice-centered relational (vcr) method as a 'listener's guide' (see Brown 1997; also Code 1993 and Byrne, Canavan and Millar 2009) to ask and answer three interrelated questions: i) who is speaking, under what circumstances, ii) who is listening, and iii) what is the nature of the listener–speaker relationship?

Literally, *the voice of participants* lies at the heart of the vcr-method and it directs the researcher to exercise intensive listening, not only during the interview, but also at the time of reading and interpreting the interview transcripts. There are two reasons why I consider this particular method to have explanatory power in the analysis: first, because of its analytical potential to examine agency, which is at the core of my study, and second, because of the reflexive element that is embedded in it. Applying the method emphasises, in this case, the voices of ten highly educated Tanzanian women and highlights listening to how they speak about themselves *before* I talk about them. By listening to 'who is speaking, under what circumstances', the method provides a sense of agency and social position, as expressed by the participants, and it recognises the importance of the social context, which are essential elements of my study. Furthermore, the method stresses the participants in plural, referring particularly to the role of the researcher both in the research process and in knowledge construction. The vcr-method was first conceived within the psychological paradigm, but it has been applied and developed further in different disciplines. I have been inspired by the sociologically oriented work of Mauthner and Doucet (1998, 2003), and have built my analysis on their model, which includes four differently focused readings of interview transcripts.

The first reading/'listening' is two-sided: to begin with, I listened to the *overall story* asking the question, 'Who is telling what story?', in an attempt to capture main events, actors and relationships, repeated words, themes, key images and possible inconsistencies. Alongside the narrative, I asked the question, 'Who is listening?', and traced and documented my *own responses* to the woman and her story. The reflexive element of the first listening aims to locate the researcher socially and theoretically in relation to the narrator, and encourages the explicit documentation of how interpretations of narratives are made (Byrne, Canavan and Millar 2009). Mauthner and Doucet (2003) consider our analytical (re)presentations to be always self-presentations as well. I agree and appreciate this stance. Similarly, I am consistent with them in understanding research accounts as 'one story among an infinite number of possible stories', and wish to highlight the importance of stories, many stories, as presented in the Introduction.

The purpose of the second vcr-listening is to identify how the narrator *represents herself*, while the third reading traces the *relationships between her and others*, and the consequences of these relationships. At the time of second and third reading/ 'listening', I pinpointed when and how the women used 'I', 'me', 'us', 'we', 'you' and 'they'. This is the point of the analysis to identify the experienced and perceived agency and social location. In her study on femininity and white working-class girls in rural Maine, Brown (1997) highlights that much of our sense of who *we are* arises from who we believe *we are not*. For that reason, Brown focused on the girls' gossip and put-downs of their peers and siblings, aspiring exactly to find out whom they considered 'others' and why (ibid.). Accordingly, by observing how the participants in my investigation expressed and conceived themselves and others, I intended to understand what, in their opinion, made the difference to allow them to overcome educational barriers and reach the highest level of education, despite all the odds, as the majority of 'others' around them did not (third reading/'listening').

The fourth vcr-listening is the analytical stage during which the micro-level data is set in relation to the macro-level analysis. As discussed above, Tanzania, on one hand, provides the country case for my study, and contextualises the study, on the other. Accordingly, at the time of the fourth vcr-listening, I located the women's narratives within the broader social and structural setting, by which I mean the strategic priorities and practices executed within the recent education sector reforms in Tanzania.

All of the four readings/'listenings' emphasise the relational nature of the vcr-method. In my research, the fourth listening posits the subject-agency in relation to the educational establishment and societal structures, while the third and second listenings relate the subjective practitioner to more private social and familial environments. In addition to ontological relationalism, particularly at the time of the first listening, the vcr-method posits the researcher-subject in epistemic relation to the one who is researched. The cultural/reflexive turn in social and human sciences has questioned the idea of objectivity and neutrality, and increased awareness of the relational and con- structed nature of knowledge. In other words, it is well acknowledged that there is a relationship between how knowledge is acquired, organised and interpreted and what its claims are. The very ideas of relationalism and reflexivity, essential and embedded in the vcr-method, imply also the evalua- tive and ethical elements of my study, to which I return in the next section.

Methodological Premises and Reflexive Accounts – Some Remarks

To recap, the purpose of my study is to develop an understanding of the com- plexities and intersecting issues of female education and schooling in the global South, and thereby to complement the knowledge base of gender equality and equity in education, gained on the basis of the hegemonic narrative. To recap further what has been said earlier, by choosing *what* to study and *how*, the researcher can present the particular kind of story that she wants to be told and heard. For this reason I wish to explicitly expose some of the methodo- logical premises on which I base my study. Similarly, some evaluative and ethical reflections are helpful.

Epistemic and Ontological Considerations

From the *epistemological* premises point of view, by definition, my study is based on hermeneutics, representing both a generic research orientation, and a theory of interpretation and understanding. As a theory of interpretation, the hermeneutic tradition has its roots in ancient Greek philosophy and rhetoric. By simplifying the hermeneutic evolvement of hermeneutic philosophy, three ways of understanding can be differentiated: one view gives emphasis to understanding *linguistic* communication, while the other stresses provision of a *methodological* basis for the human sciences (in comparison to 'natural sci- ences'); although Heidegger (1889–1976) did not completely reject the previous two, his paradigm shifted fundamentally the focus of twentieth-century her- meneutics from interpretation to *existential* understanding (see e.g. Gadamer 1975, 1976; Ramberg and Gjesdal 2009; Ricoeur 1971, 1981). The ontological turn of the paradigm was sustained by Heidegger's student Gadamer (1900– 2002), for whom existentiality and the ontological nature of human knowledge, our ability to interpret and understand, formed the core of hermeneutics. In sociology, where my background lies, hermeneutics is strongly influenced by Gadamer's work, which emphasises language, historicity and locality, the context *par excellence*. The principle that follows from both Heideggerian and

Gadamerian thinking is that our understanding is a mode of being, not something we do but something we ontologically are; our consciousness is linguistically mediated, and always culturally and historically relative. Based on Heidegger's concepts, and corresponding with the ontological premise of comprehension, Paul Ricoeur (1913–2005) developed his version of hermeneutics. However, where Heidegger and Gadamer rejected the form of exact epistemological and methodological hermeneutics, which is where Paul Ricoeur's views lie (Ricoeur 1971, 1981; cf. Hekman 1986).

Obviously, naming only three philosophers does not do justice to the intricacies and nuances of hermeneutic thinking; rather, in a simplified way it exemplifies different adaptable positions within the paradigm dependent on the research tradition. In my study the hermeneutic tradition has had, in particular, epistemological and methodological influence. I agree with the claim made by Anthony Giddens (1987, 45–46), who suggests that 'the point of doing social research, from a practical angle, is simply to allow policy-makers to better understand the social world'. To him 'the condition of establishing a dialogical relation between researcher, policy-makers and those whose behaviour is the subject of study, is inevitably one of the main contributions social research can make to the formulation of practical policy' (ibid. 47). Consequently, upon defining the purpose of scientific investigation as the development of knowing and understanding social behaviour, hermeneutics, comprehended as a theory of interpretation, is a sound and solid alternative. In addition to positioning hermeneutics as a theory of science, hermeneutic thinking guides the overall research approach, which again has practical implications for the logic and structure of the ontology of understanding. I am referring particularly to the concept of the hermeneutic circle that represents in my study a process of methodology and a method of interpretation.

Even though Gadamer repeatedly and strongly denied any methodological intent of his hermeneutic investigation, he has undeniably offered an understanding for social and human sciences which have had methodological implications. In my study, hermeneutics is comprehended and employed as a general methodological principle of interpretation *par excellence*. I see the hermeneutic circle as an iterative process through which our scientific understanding occurs, and through which we correct our prejudices or set them aside. In hermeneutics terms, our comprehension of the unity depends on the compatibility of the constitutive elements, and vice versa, since making sense of the individual parts requires having some sense of the unity. In cases where there is no correspondence between the parts and the unity, through conversations with others, a lack of understanding or misunderstanding may be altered, and a new understanding reached. I will explain next the idea of the hermeneutic circle as it has been understood and employed in my study.

As elucidated earlier, the starting point of my study was somewhat pragmatic, and based on a particular kind of literature. In hermeneutic language, this means that my pre-understanding (fore-structure for Heidegger, or what Gadamer termed prejudice) of girls' and women's education and schooling was predominantly guided by the information provided by the official Tanzanian *policy* documents and monitoring and evaluation reports. As discussed earlier, it is rather plausible to claim that one perspective only provides an inadequate understanding of the phenomenon and needs to be complemented for a fuller comprehension. In my study, the pre-understanding was transformed by the addition of another perspective namely the (previous) *research*, and now the substance and essence of the understanding was elucidated as an interrelationship between the two parts of the unity. However, to attain a more meaningful understanding of the unity, in accordance with the aim of my study, it was critical to be still instructed by the *practice* part of the unity, which I defined as the narrated experiences and insights of Tanzanian women

Even though there is neither a starting point nor a closure in the hermeneutic interpretative process, it is not the circular reasoning, but the dialogical relation that characterises the hermeneutic construction of knowledge; the dialogue between policy – research – practice in my study. For Gadamer, the hermeneutic circle is not 'vicious', because it does not describe a methodological problem (Hekman 1986). For Titelman (1979, 187–190) the circularity is not vicious either, as it 'involves a passage from pre-conceptual understanding to the seizure of its meaning'; for him 'the hermeneutical task is to find justifiable modes through which my experience and comprehension of the phenomenon being researched can serve as a bridge or access for [...] the meaning of the phenomenon' (ibid., see also Gadamer 1975; 1976, 38). I have taken these remarks into consideration, and in line with Titelman, who transformed Ricoeur's (1971) methods into a pragmatic application, view hermeneutics as an 'argumentative discipline' and a constitutive and reflective process, through which we develop our understanding of the unity, on one hand, and transform our prejudgements, on the other.

In consequence, defining hermeneutics as an epistemological process and a generic research approach gives emphasis to the text (e.g. data transcript) under investigation, but just as importantly for the narrator(s) *and* the interpreter(s) (researcher; readers). Indeed the subjective and collaborative construction of knowledge, the connection between actions and texts, is another important implication of Ricoeur's analysis for this study. Following Gadamer's concepts, in the fusion of horizons (evolvement of hermeneutic understanding), the authors (actors) are not the exclusive determinants of the meanings of actions; rather, the meanings of texts (actions) are a product of the fusion of the horizons of both the actor and the interpreter (Hekman 1986). Consequently, my explanation is, by definition, a joint construction of knowledge produced through an interaction between respondents' accounts *and* how I make sense of these accounts – collaborative accomplishment, as depicted by Ryen (2003, 439). Certainly, this leads to contingent generalisations, and a subjective, situated, partial, developmental and modest understanding of the nature of knowledge (see Byrne, Canavan and Millar 2009; Mauthner and Doucet 1998, 2003), which will be further contemplated at the end of this chapter. What has been said implies that subjects and subjectivities are predominantly guiding my study – in line with the core idea of the capability approach, which locates normative primacy for the well-being of every human being. This does not imply, however, methodological individualism in the ontological sense, to be discussed next.

As stated, in my study the *ontological* primacy is given to neither individual human beings nor structures and systems. Instead, my investigation is based on such a methodological relationalism that draws attention to positions constituted by social relations. Consequently, apart from understanding human agency and social reality as processes under construction, this positioning embeds the ontological principle of reciprocity. In short, this means that social structures, which can be enabling and constraining, socio-cultural meanings (shared understandings, expectations and knowledge) and social institutions alike constitute both the pre-condition and the outcome of people's well-being and agency. In other words, subjectivity is constituted by social structures, but the very same structures are produced and reproduced by subjects through human action.[1] The loop between agency and structure depicts also an analogy with the evolvement of the hermeneutic understanding as previously discussed.

In my examination, the relational undertone conveys the comprehension of the construction of social reality, *and* individual well-being and agency: I am examining i) social structures and institutions that provide frames and context for human action, on the one hand, and ii) individual decision-making concerning educational choices and career, on the other; since educational agency is understood more as a becoming rather than a being, I am particularly interested in investigating iii) the processes through which girls and women make their educational choices in relation to their socio-cultural structures, that is, how they 'do gender' (Unterhalter 2007b, 90) with regard to education, in relation to their school environment and their social and familial environments. This relation and 'negotiation process' is comprehended as a mediator amalgamating the structure and agency, viewed characteristically as dichotomic in social theory.

In my study, such relations and negotiations mean challenging gendered educational hierarchies and institutions, and refer to a contestation (of the *idea of Tanzanian woman*) and a process of 'becoming something else', that is, opportunity for change. By definition, to practise human agency presumes four constitutive elements: first, there is an intentional and rational agent; second, aiming at something; third, the agent has at least two means for pursuing the goal; and finally, practising agency always takes place in a certain socio-cultural environment (see Smith and Seward 2009; Nebel and Herrera Rendon 2006; cf. Alkire 2008; Crocker & Robeyns 2010; see also Kabeer 1999). To posit these presuppositions in the examination of girls' and women's education and schooling in Tanzania, as in any context in this regard, introduces the following assumptions: to begin with, presumably girls and women are rational and intentional agents capable of reflexive decision-making regarding their educational career;[2] to continue, girls' and women's actions are not determined (in an ontological sense) but are pre-conditioned and influenced by social and cultural structures. As a consequence of contextual knowledge and understanding, girls and women do have *an idea* (Hacking 2000) of the Tanzanian woman, which in turn, is embedded with assumptions regarding the aims and means available: to be educated or not, what level of education is appropriate and reachable; is it a choice made by the girls and the women, is it just 'going with the flow', as was judged by Wema, one of the research participants, in our discussion, or according to someone else's aspirations and decisions; and are there, in fact, two or more means to be chosen from, or actually none.

As pointed out by Hacking (2000), amongst many others, *ideas* do not exist in a vacuum but inhabit a social setting. In the Tanzanian context, there is a particular kind of *an idea* of girls and women that includes also the idea of education and schooling (which will be discussed in detail in Part II). The education policy and research depict a somewhat negative image of the education and schooling of Tanzanian women, and therefore the Government of Tanzania is committed to reformulating and developing *the idea of Tanzanian woman*, which suggests that constructing educational well-being and practising individual agency means reshaping the *idea* and contesting gendered inequalities in education. In my study, this kind of parallelism gives rise to questions such as: to what extent the ten research participants are practising agency, that is, capacity and autonomy to make decisions and take actions regarding their educational choices; in what ways their being is in accordance with the 'conventional' idea; and how they aimed to become something else. As pointed out in the introduction, there are a small but growing number of Tanzanian women who have reached the highest level of education, including the participants who joined my study. In that regard, they have not practised and sustained the hegemonic idea of education and schooling of girls and women, but have become something else. This relational process of becoming is at the focus of my study and is understood to be socially constructed.

As noted provocatively by Hacking (2000) and more 'constructively' by Gubrium and Holstein (2008), since the 1960s, constructionism has become a label for everything and nothing at the same time, almost everyone considers oneself to be a constructivist, and nearly everything has been defined to be socially constructed. My examination is part of the collage of social constructivist tradition seen, however, more as a research approach rather than a social theory. My study represents moderate constructionism and does not regard all processes and practises as socially, not to mention discursively, constructed; on the contrary, I view the school environment, for instance, very much as a fact, but what is constructed is the way girls and women operate – and are capable of operating – in that particular social setting, that is, classrooms and schools at large. To put it differently, I pay explicit attention to school resources, such as desks, books and 'mathematics teachers' (conversion factors), for instance; but still, from a gender perspective, it is more relevant to look at what the girls and women do and are capable of doing with the resources available, and whether that particular kind of being and doing is determined categorically by their gender. The twofold nature of social reality is also evident in the social and familial level of the enabling environment. For example, the issues of poverty, rural/urban settlement, household-family composition etc., are understood to a considerable degree as concrete resources and facts;[3] however, their gendered impacts, if any, are socially constructed, that is, results of social forms of behaviour.

As suggested by King (2004), referring to the groundbreaking work of Durkheim, Marx and Weber, the special character of social reality is nowhere other than in social relations, and for that reason, he asserts, the focus of sociological research is in social human life and relations (compare Archer 2007). As will be discussed in detail in Part II, there are different kinds of relations and different kinds of factors at different levels of the enabling environment that has an impact on girls' and women's education and schooling in the global South. Thus, gendered educational opportunities and outcomes are interrelated and influenced by both macro- and micro-level issues within the educational establishment and in- and out-of-school factors. In my study, the focus is on a few, relevant for the purpose and interest of the research. I am particularly referring to the factors related to the school environment (supply and demand) and socio-economic and socio-cultural factors intertwined with the social and familial levels of the enabling environment (demand).

Owing to constructivist ontological positioning and resonating with the evolvement of hermeneutic understanding, I explore the practical workings of *what* is constructed and *how* the construction process unfolds, working back and forth between them. I am not overtly concerned with *why*-questions; nor have I examined to what extent the women in practising educational agency reconstruct their dispositions and social environments. However, this part of the constructivist position will be commented on in the final discussion section, where, as an example, the women themselves are appraised as role models for young girls and other women. The main reason for not addressing why-questions in the study is based on the profound assumption of the ontology of the educational agency discussed earlier. As a consequence, I have presumed that the research participants have had a reason to value (to pursue) education (Sen 1985, 1992, 1999). Regardless of some references to reasoning that concerns the importance and meaning of education (for instance the value of education in itself, to achieve (better) professional qualifications, better career prospects, a better standard of living), which will be discussed later in the context of the familial enabling environment, I have not examined the reasoning *par excellence*; instead, whatever the reason to value education has been, it has been comprehended to be embedded in the very idea of a 'rational and intentional' educational agent.

What then, is comprehended to be socially constructed? Previously, I argued that social structures, which can be socio-cultural meanings and institutions alike, social constructions and 'facts', constitute the precondition for people's agency. This applies to the school environment and social and familial environment, as briefly explained above, but also to girls' and women's self-image and their idea of themselves as individual subjects.[4] According to Schwandt (2007), the notion of the self as an entity, as an autonomous, boundaried locus of decision-making and intention, may be the way that many people make sense of who they are [in western societies];[5] from the social-constructionist perspective, this is understood to be constructed, as mediated through different forms of cultural practices and meanings – *ideas*. For instance, one of the participants in this research, Rehema, is physically impaired, and another one, Leyla, remarked that 'I'm 53, I am so old' (to be a student). Obviously, disability and age are physical facts, but the way the environments, in this case the learning environments, families, communities and society enable or constraint disabled and 'old' women to function, is a social construction.[6] On this basis, I understand girls' and women's educational pathways as facts, as functionings, which they have managed to achieve, each of them in their own particular and unique way (cf. Subrahmanian 2005). However, besides retrospective scrutiny of their achieved functionings, it is of relevance as per the interest of this research to examine the construction of girls' and women's educational career as an interplay between the opportunities and constraints within the school environment and social and familial environment (capability sets), and the *idea* regarding education and schooling that the women themselves have (agency). This means positing the subjective self in relation to these two environments.

As pointed out by Marshall, 'there is likely no area in which social constructionism has had more lasting and critical impact than in feminist work on gender' (Marshall 2008, 687). She reminds us that, in addition to the categorical division of the person, gender is a premise for hierarchy, both of them being social constructs and hence, neither natural/inevitable, nor unchangeable. That is to say that gender-based differences and hierarchies, that is, gender inequalities i) need not have existed or need not be at all as they are ii) are quite bad as they are, and iii) we would be much better off if they were done away with, or at least radically transformed (Hacking 2000). This kind of criticism of the status quo, according to Hacking, is at the core of constructionism, consistent with the three key elements of feminist theory as per Chafetz (1988, 5). This has analytical implications to analyse and assess gender inequality in education in my study. To begin with, the power of social structures is acknowledged and examined, yet, for the possibility for change to reach gender equality and equity in educational institutions, analogous to gendered identities, is given normative priority. This resonates with the presuppositions of the capability approach. Furthermore, the focus of analysis is to capture strategies for change, that take place, on one hand, at the micro-level, exhibited in women's educational pathways, and on the other hand, at the macro-level through policy priorities and strategies. Finally, the correspondence of the micro-level pathways and macro-level policies is analysed to find out to what extent these two horizons, indeed, fuse.

Despite my personal (growing) curiosity towards critical realism, the ontological premise of this study, which is based profoundly on my personal

academic thinking and orientation, is importantly influenced by the constructionist tradition pioneered by Peter Berger and Thomas Luckmann: it is
their approach to and interpretation of the social construction of reality
according to which I am primarily oriented in an ontological sense, notwithstanding their initial epistemological intention (Berger and Luckmann 1966).
The tradition to which they gave impetus is not adopted and applied in the
study as a theory as such, referring in particular to the process of socialisation;
neither does it direct the interest to symbolic and linguistic interaction.
Instead, it focuses my analytical lenses, first, on how the *language* reveals
being and becoming and provides a route to meanings (compare Gadamer 1975
and Titelman 1979), and second, towards *social everyday life and customary
practices* (see also e.g. Smith 1987). In particular the aspiration to build social
theory on the basis of cultural theory depicts my affinity with Berger and
Luckmann's thinking, followed by many more recent theorists.

As a parenthesis, for Archer (2007, 19) the interplay between structure and
agency 'firstly involves a specification of how structural and cultural powers
impinge upon agents, and secondly of how agents use their own personal
powers to act so rather than otherwise in such situations'. She points out that
both the objective impingement upon and the subjective response to it are
involved, and highlights how realists (including her) have concentrated on the
former and left the latter for less attention. This reference interestingly exemplifies the constructivist positioning of my enquiry: the two previously introduced elements are involved in this study, but differently from realist
positioning, prominence is given to subjective experiences and insights (to
social everyday life and customary practices; to act so or otherwise), that is,
for girls and women to reach higher education, the normative primacy given
to the well-being and agency of human beings as previously discussed.

Evaluative and Ethical Reflections

Apart from the philosophical and methodological premises on which my
study and this book draw, more personal, evaluative and ethical reflections
are useful. When I am presenting at seminars and conferences internationally
or in Finland, colleagues consistently enquire: Why Tanzania? Why are boys
and men not included? and I answer by telling them pretty much the same
story that was outlined in the introduction. However, behind the pragmatic and
rational reasoning, there are ethical considerations, subjectivity and evaluative
aspects alike, weighted alongside the research process which have been implied
to some extent already, and discussed and reflected on explicitly here. I focus
more on practical ethical issues at different stages of the research. The relevance
of the chosen research methodology as a whole is discussed more in the final
chapters in Part IV.

To start with: why Tanzania? Twenty years ago, Reinharz (1992) remarked
on how Western feminists were criticised for not studying Third World women
(ethnocentrism), and were criticised for doing so as outsiders to those cultures

(colonialism). She then stated how feminists studying women in cultures other than their own were criticised if they accepted that culture's way of subordinating women (misogyny), and were criticised if they repudiated the culture's subordination of women (ethnocentrism again). Overall, as she concluded, feminists doing cross-cultural research seem to confront two competing sets of ethics: respect for *women* and respect for *culture*. Consequently, each person who contemplates such research must decide where she stands. Although the rhetoric of 'Third World Woman' has been strongly criticised and contested since Mohanty's (1988) ground-breaking article, echoed, for instance, in the concept of the 'girl child' in the context of educational development (e.g. Fennell and Arnot 2009), embedding the monolithic idea of women in the global South as a homogenous group of oppressed females without agency, the points of view that Reinharz brings up are still very much valid. Besides, as Mohanty (2003) later noted, 'discursive sites for struggle' are as important as the material ones, because of the essentialist character of language. For Reinharz, a useful guidepost to posit oneself within the women/culture ethics discussion is a reflexive writing process, to which I return in a moment. First, I answer the question of why Tanzania, choosing 'the case', as the very first stage of the examination.

As explained above, at the time of deciding to undertake research in general, and to focus on Tanzania in particular, I used to work in a consultancy company. The reason for becoming more interested in the theme of education, gender and development was very much in accordance with the core of the capabilities approach (as will be presented in Part II), that is, to question the information and knowledge base used to do development, and to assess development and change. At this stage of the research, I conceived of the phenomenon to be examined as a *fact*, as a (socially constructed) problem of gender inequality in education in Tanzania, bad as it is (cf. Hacking 2000, Chafetz 1988), which the government of Tanzania is committed to tackling and eradicating (in line with international policy commitments); and the government of Finland was committed to supporting these processes. Thus, *what* to study was not so much of an ethical concern, and thereafter too, the 'education for development' ideology represented to me as somewhat self-evident, embedding both the intrinsic (basic human rights) and instrumental (benefit for individuals) values, which I agree with and value both personally and academically; however, *how* to study involved more ethical and evaluative deliberations.

To complement the dominantly macro-level and quantitatively focused information and knowledge base, a micro-level and qualitative approach was a fairly evident choice, and the indicators of well-being and agency were interesting 'proxies' of subsequent gender equality, collected from women who, presumably, had achieved well-being and exercised agency, manifested as their educational pathways. In the initial research project plans, our idea was to carry out more of a *gender* study; that is, to also incorporate the male perspective, and to include more of the social (community) level aspects of the enabling environment, but then I decided that I wanted to focus only on the women's experiences and insights, as representatives of unique educational achievements in the country, even if measured quantitatively.

The starting point for our research project was to establish an institutional collaboration between my home university and the University of Dar es Salaam (UDSM): first, framed as a Memorandum of Understanding; later, signed as an Agreement for Co-operation. Simultaneously, collaboration with Tanzanian researchers was initiated. Throughout the and the study which followed, the research was planned and implemented in liaison with Tanzanian partners. Furthermore, in order to conduct research work in the Tanzanian context, a permit from the Tanzania Commission for Science and Technology (COSTECH) is required. Being supported by the School of Education at the UDSM, the research clearance was granted to me, enabling me to carry out a study in the country. To initially get in touch with the highly educated Tanzanian women, and to be introduced to the research participants, I was supported by our Tanzanian research advisor through her contacts. I did not know, however, when I first met four of the research participants, how our research advisor had, for instance, introduced me and my work to the women, but as mentioned above, apart from the formal consents that I got from these women, I got the impression that they were curious to join the study, and *I* experienced our encounters as being very pleasant, and indeed, more like conversations. The women were interested in knowing how I intended to use their stories and we made some jokes about how I might make a Hollywood film out of the tapes, in addition to the usage of the data for my dissertation and in other research publications. I enquired if they would mind me using their photos in my study, but they were not willing to give their permission – clearly, there was a border between 'giving voice' and 'giving face' to the experiences and insights. For this reason, even though I have their permission to use their first names, I decided to use pseudonyms, just to avoid any potential inconveniences.

In accordance with the methodology that I have chosen, I wish to exercise researcher's reflexivity and some storytelling concerning *how I perceived* the first data-collection experience. The following 'picture is drawn and story told' on the basis of my research notes:

There I was, at Onnela, a compound in a residential neighbourhood in Dar es Salaam, sitting with my Tanzanian counterparts in a classroom used for the purpose of 'School for Finnish children'. I was excited. I hadn't met my research participants earlier and I didn't know what they knew about me, about my work, and about the purpose for which they were actually there sitting next to me. The video recorder was running. I introduced myself, welcomed them, explained to them about my educational and professional background, about our research project and my doctoral studies, and about the intended interview to be carried out. They then interrupted me and started to ask questions about my nearly two-year old son and husband, whom I had just briefly introduced to the women, when running into them in the Onnela garden. They wanted to know if I had only one child, if I intended to have more, whether my husband really took care of our son when I was working, what he was doing for a living, and whether we were both well and truly working and taking care of our son at home. For me, this was the first lesson to be learned: instead of only being a highly educated white Finnish woman in my thirties, someone pursuing a PhD degree, asking odd and probably difficult questions, and speaking a rather peculiar language, it was much more important and relevant for the women to know that I was also a mother and a wife – in that order; that instead of educational choices, opportunities and constraints, which we were supposed to discuss later, I was also a private person trying to balance and make decisions concerning professional aspirations and my private life. Discussions went on, I felt comfortable and at ease, and got the impression that they felt the same. We discussed their educational life stories. I become fascinated and I was impressed!

After a couple of days, I met a few of them again. I figured we could get started right away and wondered if there would be any need for small talk before we began. But, Amana and Leyla began by telling me how they had seen my husband and son at the university campus, and that my husband had been carrying our son in a kid's carrier, and how they had had a cola at the campus cafeteria; they exclaimed that my husband and son had amused pretty much the whole campus. This, again, was another lesson for me. First things first! It seemed not only important, but fair to them, for me to mutually bring my experiences and life closer to them, as they were sharing their stories and lives with me. Isn't this similar to what is done at the canteen at work with women colleagues? Aren't the mundane and simple conversations the most natural and logical way to try to at least diminish the gap between our lived and experienced realities, since we do not even consider entirely closing it?

In being reflexive, as remarked on by Adkins (2002), there is a disturbing tendency that the 'narration of the self', rather than reflexivity *per se*, is given authority in the research. I wish to avoid this, but yet, as emphasised by Lumsden (2009), Mauthner and Doucet (2003) and O'Reilly (2005), and in line with the idea of the 'reflexive turn', I think that the personal biographies, in addition to the academic background, importantly constitute the sources of knowledge; thus, it is important to situate oneself within the 'texts' under investigation through autobiographical accounts and personal narratives. Axinn and Pearce (2006, 8) discuss reflexivity in deliberating on the use of the participant observation method. They agree that participatory observation may be useful in providing researchers with the opportunity to put themselves 'in the shoes' of the people they study and use introspection as a tool; however, they claim, researchers can never fully fill these shoes. For example, 'when an American researcher goes to Nepal and transplants rice, he or she is unlikely to ever feel exactly like someone who does it every year and who knows that if the crop fails he or she will not eat in the coming year' (ibid.). I agree with this, and I am more than willing to admit that I can never really understand such a profound issue as the poverty in the country, and its consequences for the everyday life and arrangements of the Tanzanian people, and women in particular. Neither can I ever truly understand, let alone agree with, the very gendered *ideas of Tanzanian woman*, nor of the categorical conception of a woman in general in that regard. Still, I think, in my case, for example, as may be inferred from the above narrative, I was married and we had our first child at the time of the first interviews, and being a mother strongly impacted on my experiences in the field, on listening to and interpreting the experiences and insights of other women who were also mothers, wives, professionals and students, and undoubtedly, as may be inferred from the narrative above, my position definitely influenced how the research participants perceived me, and how we related to each other.

As pointed out by Byrne, Canavan and Millar (2009; see also e.g. Laws, Harper and Marcus 2003), there is a great deal of literature on participatory approaches in planning and implementing research. However, they continue, much less is discussed concerning the process and practices of participatory interpretation and analysis (ibid.) This would have been an interesting idea that was worthwhile adopting, as I was operating in a context that differed to such a great extent from my own socio-cultural and socio-economic environments from the perspectives of both reliability and research ethics to check if I had understood correctly what the women had told me. And wouldn't it have been just fair to provide an opportunity to comment. I am not suggesting that the research participants are to be held responsible for the correctness of my inferences and claims; instead, I think that it would just do justice to the research participants, in any study, who have told a stranger about issues that concern their private lives. The focus of Mauthner and Doucet's (2002, 2003) methodological concern is also in the analytical and interpretative stage of research processes, but I think that the ethical considerations, the subjectivity and evaluative aspects which they emphasise, are also applicable to other research practices. However, because I did not adapt any kind of participatory analysis, I have tried to pay attention to my own biases and prejudices, which I had, as I found out, in order to avoid incorporating these into my analyses (e.g. Burawoy 1991). For example, Leyla told me how they used to grow coffee and how that type of farming was something that men typically took care of and were in charge of; she also described how it was customary for girls to collect grass for the cows and for boys to feed the goats by climbing the trees to cut the branches for food. To me these were, at first hearing, evident examples of discrimination and a (negative) gendered division of labour: the coffee as a cash crop and *naturally, of course*, male business; and the climbing of the trees seen as just inappropriate for girls, it being presumed that they were not able to do that anyway. However, as Leyla explained, many pesticides are used in growing the coffee, and often the women who work in the family fields do not have time to go and wash themselves before they breast-feed their babies; for this reason, she said, women do not really farm coffee, no matter if the business idea also exists. She also perceived that their families were protecting their daughters in not allowing them to climb high up in the trees, and although one may question if there is, behind that, still the idea of what girls can do in comparison to the boys, an alternative is just to agree that the family members truly and honestly wanted to look after the girls and did not harbour any form of discrimination.

Obviously, my kind of a way of hearing and interpreting narratives is packed with influences that are based on my personal history and position, institutional and interpersonal contexts, academic background and the kind of theoretical and methodological literature on which I essentially base my scholarly thinking, and of course pragmatic issues. The voice-centered relational method is *one tool* with which to try to exercise reflexivity, to explicate the assumptions and commitments (personal and/or saturated in the literature) and to incorporate reflexive observations and accounts into the analysis; that is, to validate the interpretations by being able to demonstrate how they were reached (Mauthner and Doucet 2003; also Mason 2002). However, I wish to avoid being naïve, since there are definitely limitations on how reflexive one can be, and how far one can know and understand what shapes the research at different stages at the time of conducting it. As noted by Mauthner and Doucet (2003, 419), 'it is only sensible to give prospect for hindsight, since 'there may be limits to how reflexive we can be, and how far we can know and understand what shapes our research at the time of conducting it, given that these influences may only become apparent once we have left the research behind and moved on in our personal and academic lives'. Accordingly, it is difficult to assess to what extent I have used the 'badges', referring to the acknowledgement of the importance and influence of 'characteristics such as gender, ethnicity, class, sexuality and geographical location', having however, very little impact on the actual research or the interpretative process, although they are meant to represent one's respect for difference (ibid.) I also wish to avoid the tendency to romanticise the research participants' voices, no matter how unique their educational pathways or how marvellous I perceive these women to be. I do not presume this research to bring any particular changes into the women's lives, regardless of the fact that I was really happy and privileged to receive feedback at our last group session from Amana, thanking me and describing how: 'Through your questions and our discussions, we have also learned: we have learned that we have done something, that we are somebody.' However, I am unwilling and rather reluctant to apply the empowerment discourse in the study, which to me implies that previously, the women were not empowered. However, it is of great importance to be aware that every research activity is an intervention, no matter if such a seemingly small and insignificant act as a conversation with another person, and may bring along intentional consequences such as complemented information and knowledge base, a developed understanding of girls' and women's education and schooling in the Tanzanian context and a PhD dissertation and degree, or unintentional consequences, such as the aspect of empowerment that was explicit in Amana's deliberation regarding her beginning to think differently of her own accomplishments and life.

What is said above suggests that subjects and subjectivities are at the core of this examination, and are not considered as limitations on the validity, reliability or objectivity of the research; quite the contrary, they are understood as an integral part of the relational process of knowledge construction, resonating and amplifying the epistemic positioning of the research. In consequence, subjective, situated, partial, developmental and modest understanding characterises the nature of the knowledge, attained as a 'collaborative accomplishment' (Ryen 2003) between the research participants' accounts *and* how I make sense of these accounts. In the next three chapters, we hear subjective accounts, by definition, as research participants' experiences and insights concerning the degree to which their schools and social and familial environments are enabling.

Notes

1 Margaret S. Archer (2007, 17) points out that one of the few propositions upon which social theorists agree is the truism 'no people; no society'. She continues that no one seriously maintains that 'society is like people'; instead, it remains different in kind from its constitutive elements even as its being is understood as no more than the aggregate effect of people's doings and patterns.
2 Margaret Archer (2000, 2003, 2007, 2010) for instance, has argued for the cruciality of self-awareness and reflexivity. Also, seeing human beings as active agents of their own lives, capable of reasoning, and not as passive recipients of various (development policy) initiatives, is foundationally embedded in the capability approach, as will be discussed in Part II.
3 This position is contrary to that of Gadamer, who would describe it as 'absolutely absurd' (because it reaches outside the scope of language), and close to that of Habermas, who sees a position outside socially constructed reality. Habermas posits a realm of objectivity, from which what he sees to be the subjective, ideological conceptions of social reality can be assessed and criticised; he strongly emphasises the determining role of history and culture, but offers still a theory stepping outside the social reality created by these influences (Hekman 1986, 37–38).
4 'Adaptive preferences' in the capabilities approach are discussed by e.g. Nussbaum, 2000; Unterhalter 2007b, 2012; Walker 2007; Watts and Bridges 2006; see also e.g. Kabeer 1999.
5 Akande (2009) and Hart and Poole (1995), to name just a few, point out the importance and impact of cross-cultural differences regarding the dimension of individualism–collectivism, to be considered both as a methodological and analytical issue.
6 Chege and Arnot (2012) discuss schools as gendered institutions (linked into local construction of gender and to gender segregations within poverty); see also Arnot et al. 2012a on domestic gender relations in rural Ghana and India, and Arnot, Chege and Wawire 2012b on gendered constructions of citizenship in Kenya.

Bibliography

Adkins, L. 2002. *Revisions: Gender and Sexuality in Late Modernity.* Philadelphia: Open University Press.
Akande, A. 2009. Comparing Social Behaviour Across Culture and Nations: The 'What' and 'Why' Questions. *Social Indicators Research* 92, no. 3: 591–608.
Alkire, S. 2008. *Concepts and Measures of Agency.* Vol. 9 of *OPHI Working Papers.* Oxford: OPHI.

Archer, M.S. 2000. *Being Human: The Problem of Agency.* Cambridge, UK: Cambridge University Press.

Archer, M.S. 2003. *Structure, Agency and the Internal Conversation.* Cambridge, UK: Cambridge University Press.

Archer, M.S. 2007. The Ontological Status of Subjectivity: The Missing Link Between Structure and Agency. In *Contributions to Social Ontology,* ed. C. Lawson, J. Latsis, and N. Martins, 17–31. New York: Routledge.

Archer, M.S. 2010. Introduction: the Reflexive Re-turn. In *Conversations About Reflexivity,* 1–13. Abingdon, Oxon: Routledge.

Arnot, M., F.N. Chege, and V. Wawire. 2012b. Gendered Constructions of Citizenship: Young Kenyans' Negotiations of Rights Discourses. *Comparative Education* 48, no. 1: 87–102.

Arnot, M., R. Jeffery, L. Casely-Hayford, and C. Noronha. 2012a. Schooling and Domestic Transitions: Shifting Gender Relations and Female Agency in Rural Ghana and India. *Comparative Education* 48, no. 2: 181–194.

Axinn, W.G., and L.D. Pearce. 2006. *Mixed Method Data Collection Strategies.* Cambridge, UK: Cambridge University Press.

Berger, P. and T. Luckmann. 1966. *The Social Construction of Reality: A Treatise in the Sociology of Knowledge.* New York: Anchor Books.

Brown, L.M. 1997. Performing Femininities: Listening to White Working-Class Girls in Rural Maine. *Journal of Social Issues* 54, no. 4: 683–701.

Bujra, J. 2006. Lost in Translation? The Use of Interpreters in Fieldwork. In *Doing Development Research,* ed. V. Desai and R.B. Potter, 171–179. London: Sage.

Burawoy, M., ed. 1991. *Ethnography Unbound: Power and Resistance in the Modern Metropolis.* Berkeley: University of California Press.

Byrne, A., J. Canavan, and M. Millar. 2009. Participatory Research and the Voice-Centered Relational Method of Data Analysis: Is It Worth It? *International Journal of Social Research Methodology* 12, no. 1: 67–77.

Chafetz, J.S. 1988. *Feminist Sociology: An Overview of Contemporary Theories.* Itasca: Peacock.

Chege, F.N., and M. Arnot. 2012. The Gender – Education – Poverty Nexus: Kenyan Youth's Perspective on Being Young, Gendered and Poor. *Comparative Education* 48, no. 2: 195–209.

Code, L. 1993. Taking Subjectivity into Account. In *Feminist Epistemologies,* ed. L. Alcoff and E. Potter, 15–48. New York and London: Routledge.

Crocker, D.A., and I. Robeyns. 2010. Capability and Agency. In *Amartya Sen,* ed. C.W. Morris, 60–90. Cambridge, UK: Cambridge University Press.

Fennell, S., and M. Arnot. 2009. Decentring Hegemonic Gender Theory: The Implications for Educational Research. RECOUP Working Paper No.21. London, UK: DFID.

Gadamer, H.-G. 1975. *Truth and Method.* New York: Crossroads.

Gadamer, H.-G. 1976. *Philosophical hermeneutics.* Los Angeles: University of California Press.

Giddens, A. 1987. *Social Theory and Modern Sociology.* Cambridge, UK: Polity Press.

Gubrium, J.F., and J.A. Holstein, eds. 2008. The Constructionist Mosaic. In *Handbook of Constructionist Research,* 3–12. New York: The Guilford Press.

Hacking, I. 2000. *The Social Construction of What?* Cambridge, MA: Harvard University Press.

Hart, I., and G.D. Poole. 1995. Individualism and Collectivism as Considerations in Cross-Cultural Health Research. *Journal of Social Psychology* 135, no. 1: 97–99.

Hekman, S.J. 1986. *Hermeneutics and the Sociology of Knowledge*. Cambridge, UK: Polity Press.

Holstein, J.A., and J.F. Gubrium. 2003. *Introduction. Inside Interviewing: New Lenses, New Concerns*. Thousand Oaks, CA: Sage.

Kabeer, N. 1999. Resources, Agency, Achievement: Reflections on the Measurement of Women's Empowerment. *Development and Change* 30, no. 3: 435–464.

King, A. 2004. *The Structure of Social Theory*. London: Taylor & Francis.

Laws, S., C. Harper, and R. Marcus. 2003. *Research for Development*. London: Sage.

Lumsden, K. 2009. 'Don't Ask a Woman to Do Another Woman's Job': Gendered Interactions and the Emotional Ethnographer. *Sociology* 43, no. 3: 497–513.

Marshall, B.L. 2008. Feminism and Constructionism. In *Handbook of Constructionist Research*, 687–700. New York: The Guilford Press. Mason, J. 2002. *Qualitative Researching* (2nd ed.) London: Sage.

Mauthner, M., M. Birch, J. Jessop, and T. Miller, eds. 2002. *Ethics in Qualitative Research*. London: Sage.

Mauthner, N., and A. Doucet. 1998. Reflections on a Voice-Centered Relational Method: Analysing Maternal and Domestic Voices. In *Feminist Dilemmas in Qualitative Research: Private Lives and Public Texts*, ed. J. Ribbens and R. Edwards, 119–146. London: Sage.

Mauthner, N., and A. Doucet. 2003. Reflexive Accounts and Accounts of Reflexivity in Qualitative Data Analysis. *Sociology* 37, no. 3: 412–431.

Mohanty, C.T. 1988. Under Western Eyes: Feminist Scholarship and Colonial Discourses. *Feminist Review* 30: 61–88.

Mohanty, C.T. 2003. Cartographies of Struggle: Third World Women and the Politics of Feminism. In *Feminism Without Borders: Decolonising Theory, Practicing Solidarity*, ed. C.T. Mohanty, 43–84. Durham, NC: Duke Univesrity Press.

Moser, A. 2007. *Gender and Indicator: Overview Report*. Brighton, UK: Institute of Development Studies.

Nebel, M., and. T. Herrera Rendon. 2006. A Hermeneutic of Amartya Sen's Concept of Capability. *International Journal of Social Economics* 33, no. 10: 710–722.

Nussbaum, M. 2000. *Women and Human Development: The Capabilities Approach*. Cambridge, UK: Cambridge University Press.

Odaga, A. and W. Heneveld. 1995. Girls and School in Sub-Saharan Africa, From Analysis to Action. World Bank Technical Paper No. 298. Washington D.C.: World Bank.

O'Reilly, K. 2005. *Ethnographic Methods*. London: Routledge.

Packer, M.J. 2011. *The Science of Qualitative Research*. New York: Cambridge University Press.

Pietilä, I. 2010. Vieraskielisten haastattelujen analyysi ja raportointi. In *Haastattelun analyysi*, ed. J. Ruusuvuori, P. Nikander, and M. Hyvärinen, 411–431. Tampere, Finland: Vastapaino.

Posti-Ahokas, H., and M.-A. Okkolin. 2015. Enabling and Constraining Family: Young Women Building Their Educational Paths in Tanzania. *International Journal of Community, Work and Family*. http://dx.doi.org/10.1080/13668803.2015.1047737.

Ramberg, B., and K. Gjesdal. 2009. Hermeneutics. *Stanford Encyclopedia of Philosophy (Summer 2009 edition)*, May 28, 2010.

Reinharz, S. 1992. *Feminist Methods in Social Research*. Oxford: Oxford University Press.

Reinharz, S., and S.E. Chase. 2003. Interviewing Women. In *Introduction. Inside Interviewing: New Lenses, New Concerns*, ed. J.A. Holstein and J.F. Gubrium, 73–90. Thousand Oaks, CA: Sage.

Ricoeur, P. 1971. The Model of the Text: Meaningful Action Considered as a Text. *Social Research* 28: 529–563.

Ricoeur, P. 1981. *Hermeneutics and the Human Sciences. Essays on Language, Action and Interpretation.* Cambridge, UK: Cambridge University Press.

Ryen, A. 2003. Cross-Cultural Interviewing. In *Inside Interviewing: New Lenses, New Concerns*, ed. J.A. Holstein and J.F. Gubrium, 429–448. Thousand Oaks, CA: Sage.

Schwandt, T.A. 2007. Hermeneutic Circle. *The SAGE Dictionary of Qualitative Inquiry.* 3rd ed. Thousand Oaks, CA: Sage.

Sen, A. 1985. Well-being, Agency and Freedom: The Dewey Lectures 1984. *Journal of Philosophy* 82, no. 4: 169–221.

Sen, A. 1992. *Inequality Re-examined.* Oxford: Oxford University Press.

Sen, A. 1999. *Development as Freedom.* Oxford: Oxford University Press.

Smith, D. E. 1987. *The Everyday World as Problematic: A Feminist Sociology.* Boston: North-eastern University Press.

Smith, M.L., and C. Seward. 2009. The Relational Ontology of Amartya Sen's Capability Approach: Incorporating Social and Individual Causes. *Journal of Human Development and Capabilities* 10, no. 2: 213–235.

Subrahmanian, R. 2005. Gender Equality in Education: Definitions and Measurements. *International Journal of Educational Development* 25: 395–407.

Titelman, P. 1979. Some Implications of Ricoeur's Conception for Phenomenological Psychology. *Duquesne Studies in Phenomenological Psychology* 3: 182–192.

Unterhalter, E. 2003b. The Capabilities Approach and Gendered Education: an Examination of South African Complexities. *Theory and Research in Education* 1, no. 1: 7–22.

Unterhalter, E. 2007a. *Gender, Schooling and Global Social Justice.* London: Routledge.

Unterhalter, E. 2007b. Gender Equality, Education, and the Capability Approach. In *Amartya Sen's Capability Approach and Social Justice in Education*, ed. M. Walker and E. Unterhalter, 87–107. New York: Palgrave Macmillan.

Unterhalter, E. 2012. Inequalities, Capabilities and Poverty in Four African Countries: Girls' Voice, Schooling, and Strategies for Institutional Change. *Cambridge Journal of Education* 42: 307–325.

Walker, M. 2007. Selecting Capabilities for Gender Equality in Education. In *Amartya Sen's Capability Approach and Social Justice in Education*, ed. M. Walker and E. Unterhalter, 177–195. New York: Palgrave Macmillan.

Watts, M., and D. Bridges. 2006. Enhancing Students' Capabilities? UK Higher Education and the Widening Participation Agenda. In *Transforming Unjust Structures*, ed. S. Deneulin, M. Nebel, and N. Sagovsky, 143–160. Dordrecht, Netherlands: Springer.

Part II

Education, Gender and Development

5 Research on Education, Gender and Development

As pointed out in Chapter 2, female students' education is not only about schooling, nor is it only influenced by school-related factors; instead, it is impacted on by political and institutional factors, socio-economic and socio-cultural factors. The governmental macro-level factors affect the supply for women's participation in educational systems, while the micro-level factors are usually associated with the demand; the school-related factors seem to affect both supply and demand for female education.

As stated by Odaga and Heneveld (1995), *in-school* and *out-of-school* factors are interrelated and should be viewed as a unitary concern; hence, they suggest, home, community, school and governmental levels of the enabling environment should be addressed simultaneously. Most often, however, educational reforms and initiatives to reduce the gender inequalities in education are implemented in a substantial way in-school and are targeted at the school-related factors – in accordance with the mandate and scope of educational policy. However, based on the knowledge and understanding of the complex and intersecting factors that have an impact on female education, this singular emphasis is insufficient and needs to be complemented essentially with the factors attached to the social and familial levels of the enabling environment.

The encompassing review on 'Girls and Schools in sub-Saharan Africa' by Odaga and Heneveld in 1995, which covers many country case studies, including Tanzania, and considers various topics and themes which influence girls' and women's education and schooling, constitutes one of the cornerstones of my study. Their investigation more than 20 years ago is complemented and mirrored by more recent studies encompassing the gendered patterns regarding access to education, attainment and performance. For instance, an extensive literature review on girls' education and gender equality has been conducted by Unterhalter and her colleagues (Unterhalter et al. 2014). Based on the categorisation of factors having an impact on gender inequities in education by Odaga and Heneveld, and following their comprehensive and detailed list of barriers and problems for girls' and women's education and schooling in the developing countries, I have organised the core themes and contents for my study. 'Political and institutional factors' to contextualise the study, and to carry out a situational analysis of the education sector in Tanzania to learn what has been done, what has been achieved, and what remains a challenge, will be presented in the next chapter. In this section the focus is on school- and home- and community-related issues (see the enabling environment in Figure 2.1). Thus, in the following sections, school-related and socio-economic and socio-cultural factors, which significantly influence parental and familial decision-making in terms of whether to invest in female education, are discussed. The somewhat exhaustive list of factors that influence female education and schooling, which are related to the *social and familial* levels of the environment and *school environment*, is summarised in Table 5.1 and analysed in the two following sections.

School Environment

Odaga and Heneveld (1995) concluded their review of the school-related factors affecting female students' education with two statements: first, they remarked on how through exclusion, avoidance and marginalisation, the schools reflect and

Table 5.1 Factors Influencing Female Education and Schooling.

Factors Related to the Social and Familial Environment	Factors Related to the School Environment
Poverty: direct costs of schooling high opportunity costs	Poverty: (prohibitive) costs of schooling
Level of parental education: low socio-economic status, social class rural/urban residence	Schools environment: lack of sex-segregated sanitary facilities lack of basic amenities (water, electricity, meals) lack of learning materials
The economic value of girls: need for girls' input in income-generation need for girls' input in household chores	The learning environment: teacher attitudes and pedagogy corporal punishment and strict discipline biased and stereotyping curricula, textbooks and subject choices lack of female teachers and governors (role models; gender awareness among teachers)
Parental/familial investment behaviour: reluctance to invest in education low expectations of the economic value of schooling for girls	
Familial perceptions of the irrelevance of schooling for girls	
Initiation (ceremonies) Early marriage and bride/wealth systems Religion	Safety issues: type of school distance to school sexual harassment pregnancy
Gender socialisation: cultural beliefs, values and norms roles which society expects women to fulfil	Girls' and women's own expectations and motivation
Limited employment prospects	Girls' and women's high levels of repetition, dropout and failure
HIV/Aids	

promote society's low expectations of girls; second, according to their under-standing, the recent studies clearly indicate various forms of discrimination that limit female students' academic potential, affecting their access to schools, their learning achievements and level of attainment. More recently, the issues of access, transitions and equity in education in the context of sub-Saharan Africa have been discussed, for instance, by Lewin (2009), identifying five different educational patterns (with regard to problems and possibilities), Ampiah and Adu-Yeboah (2009), examining school dropouts in northern Ghana, and Motala, Dieltiens and Sayed (2009), discussing the impact of physical access to schooling, in a special issue of *Comparative Education*. One of the most prominent factors is *poverty*, discussed and polemised by Odaga and Heneveld (1995), Chege and Arnot (2012), Colclough et al. (2000, 2003), Leach (2003a), Lugg, Morley and Leach (2007), Vavrus (2002a), Wedgwood (2005) and reported lately, for instance, by UNICEF (2012b), linked, on the one hand, to the costs of schooling, and on the other, to the poverty of schools.

For a poor country (such as Tanzania) and for poor families, education is a massive investment. In Tanzania, the government has decided to ease the household-level burden by abolishing school fees and other mandatory parental costs from primary education, but despite this 'free' education, other direct costs of schooling still include uniforms, textbooks and exercise books. Additionally, as noted by Colclough, Rose and Tembon (2000), schooling costs might comprise other expenditures, either in cash (such as payments for lunch) or in kind. Furthermore, parents might be obliged to contribute to the construction and maintenance of school buildings. Hence, the school-related factors, i.e. the direct costs, may prohibit families from sending their children to school, and when resources are scarce, the girls of the households tend to be the excluded ones.

According to Odaga and Heneveld (1995; also Colclough, Rose and Tembon 2000; and Lewin 2007, 2009), the literature clearly demonstrates that the lack of resources to cover the direct and opportunity costs of education is a major constraint on girls' education. The opportunity costs will be analysed more in the context of families and communities. However, another aspect relating to direct costs, as pointed out by Odaga and Heneveld (ibid.), is that there are different country experiences concerning the type of institution (private or government funded), particularly where the educational outcomes are concerned. In Tanzania, students performed better in state-subsidised schools, because private secondary schools had a higher pupil-to-teacher ratio and a larger proportion of low-quality teachers than public schools; however, in numbers, more girls than boys were enrolled in private schools (Mbilinyi and Mbughuni 1991). Similar kinds of experiences were reported from Kenya when comparing government and community schools (Odaga and Heneveld 1995). As we will learn later, the majority of the participants in this research attended governmental secondary schools (with a scholarship), but some of the participants went to private schools, which they considered to be very good; and to get access to such high quality schools required a great financial effort from their parents (cf. Vavrus 2002a). All in all, from the cost perspective, as pointed out earlier in relation to (growing) regional disparities, it goes without saying that the division of households on the basis of income and 'affording' does not promote either social equality or gender equality in education.

Apart from the costs of schooling, poverty is apparent in the lack of schools, classrooms, equipment and teaching and learning materials. Often, the basic amenities such as water, electricity and school meals are inadequate or totally missing, as evidenced in numerous studies covered in the study by Odaga and Heneveld (1995). The fact that there are no schools close by, or that the quality of the ones which are accessible is poor, may result in the parents deciding not to enrol their children in school or in withdrawing some, if not all, of their children from school. According to policy monitoring and evaluation in Tanzania, the construction of schools was found to be one of the major successes of the Primary Education Development Plan I (PEDP). Hence, a shortage of schools is not a major concern in the country today, because basically every village has its own primary school. However, the quality of the construction work was reported to be somewhat questionable, classrooms overcrowded and there were simply not enough desks and chairs for all of the students. This may induce harassing behaviour and be particularly unfavourable for female students, as noted by Colclough and his colleagues (2000). Similarly, the distance to school may increase parental apprehension concerning the safety of their daughters and may be a reason to hold them back from school, as was found by Unterhalter (2003b) when investigating the complexities with regards to gender and education in South Africa. The distance to schools was also touched on by the participants in this research, especially in terms of the fear of being late and being punished. Additionally, the aspect of in-school and out-of-school harassment was covered in their stories, as we will hear later.

The Tanzanian civil society organisations (CSOs) have been rather critical towards the achievements of the recent education sector reforms, and, for instance, HakiElimu (established and well-recognised NGO operating in the field of education development and democracy) and FAWE Tanzania (the Forum for African Women Educationalists) have raised the issue of the inadequacy of basic amenities such as water and toilet facilities. From a gender perspective, there are three aspects to be considered with regard to gender-segregated sanitary facilities, or to be more precise, the lack of them. First, it is only decent to fulfil the minimum requirement of protection of privacy at schools for female and male students alike. Second, for female students, it is pivotal to have proper and separate sanitation facilities after they reach puberty, because a lack of an appropriate physical environment may lead them to remain at home, and returning to school is not that easy. This difficulty is linked to the third aspect, which is the lowering of the level of performance caused by being absent from classes, particularly in mathematics and science-related subjects. These aspects have been pointed out, for example, by Brock and

Cammish (1997), in a study identifying factors that affect female participation in education in seven developing countries (including Cameroon and Sierra Leone in Africa), by Colclough et al. (2003) in examining Guinea and Ethiopia, and by the FAWE Tanzania (2001, 2003, 2006).

The allocation of funds to construct, maintain and improve school infrastructure was one of the strategies adapted by the PEDP I, and alongside this, one of the quality improvement components was to purchase and provide a sufficient amount of learning materials. The availability of textbooks to students was identified as a key determinant for student performance in studies reviewed by Odaga and Heneveld (1995). For all that, there are somewhat critical questions posed by HakiElimu (2005, 2010) in assessing the progress of the PEDP I and particularly the PEDP II when it comes to the availability of learning materials, echoing the dilemma between quantity and quality. Apart from the sufficiency of learning materials, the image that female students face in school (textbooks, subject choices and curricula) is rather biased and stereotyping, yet, powerful in shaping their self-perceptions and views of themselves. As notified by Odaga and Heneveld (1995), several studies stress the need for more positive and accurate representations of women and their contribution to the economy and society at large. They (ibid.) point out how girls and women are nearly invisible and excluded in the books, and how a clear gender preference is given to males in the texts. Similarly, the studies discuss the limited options available for female students within the educational system, manifested in a strong bias in subject choices and mirroring *the idea* of female subjects. Finally, the lack of demand for formal education for females is often related to the curricula, understood as irrelevant to the daily lives and practical skills for the future employment of students, and especially of girls (Odaga and Heneveld 1995, Colclough, Rose and Tembon 2000) Due to socio-cultural factors, including the idea of perceived and preferred career possibilities, female students tend to opt for subjects that steer them into education, health and administrative employment. These kinds of experiences are also found in the research participants' stories.

To move from learning materials and the (physical) school environment to learning as such directs us inevitably to teaching and teachers. The research literature suggests that within the school environment, teachers' attitudes, behaviour and teaching practices have perhaps the most significant implications for students', notably female students', persistence and academic achievement and attainment (Odaga and Heneveld 1995; Barrett 2007; Vohua, Kiragu and Warrington 2012; see also Räsänen 2007 discussing teaching as an ethical profession). Acknowledging this, and to ensure the improvement of the provision of education, the government of Tanzania placed a high value on developing the quality of teachers and teaching. As pointed out by Tao (2013), research has commonly viewed teachers either as the cause of poor education quality (evidenced in absenteeism, rote teaching and withholding content) or as the victims of faulty school environments (including working and housing conditions). Due to poverty, governments are unable to pay teachers' salaries regularly, and funds for running schools are disbursed intermittently; this, in turn, results in teacher's absenteeism and lack of motivation, and compels them to look for alternative sources of income, for instance, in the form of private tuition. As a matter of fact, the tuition system and practices were noted by Wedgwood (2005) in relation to the quality of education, echoing the concern about this grey area of teaching and the particular anxiety regarding purposefully poor teaching at schools in the hope of gaining in terms of private tuition (i.e. extra income). It is obvious that the quantity and quality of time that teachers spend on teaching impact on and correlate with students' performance and levels of attainment.

To tackle the issues of low motivation, poor pedagogical skills and the deficient subject knowledge of the teachers, two key strategies were implemented in the country: teacher education reform through revised teacher training curricula (see e.g. Hardman, Abd-Kadir and Tibuhinda 2012; Vavrus 2009) and improvement of the teaching conditions (e.g. Sumra 2005; cf. Towse et al. 2002; cf. also Lwaitama, Mtalo and Mboma 2001). However, despite the development of instructional and pedagogical practices in theory (including 'gender sensitivity' and 'child-friendliness'), in reality students face poorly skilled and poorly motivated teachers, who use corporal punishment and strict discipline as a method of learning, evidenced in a number of cases reviewed by Odaga and Heneveld (1995), and as found in studies by Bennell and Mukyanuzi (2005), Towse and his colleagues (2002) and Barrett (2007), to name a few, and also in the narratives of the participants in this research. However, to avoid sketching a totally pessimistic image of the quality of teaching, it needs to be noted that there are excellent, hardworking and dedicated teachers in Tanzania 'who are trying their best', as stated by one of the participants in this research (see also Vavrus 2009; cf. Malangalila in Towse et al. 2002, 649), similar to findings by Vohua and her team (2012) from Kenya – 'against the odds', as they framed it.

To sum up, teachers' competence, experience and commitment are essential for both female and male students' successful learning. To observe the gender aspect further, Odaga and Heneveld (1995) found evidence that both male and female teachers believe boys to be academically superior to girls; in addition, some classroom observations revealed how teachers' paid more attention to boys than to girls, or how they completely ignored the girls. In the Tanzanian context, Thomas and Rugambwa (2011) have paid attention to classroom discourses and teachers' understanding of gender, and showed how both 'gender as equity' and 'gender as power relations' manifest and interact in schools. Hence, against the common presupposition that the quality of teacher–student interaction is negative and discouraging towards female students, there is evidence of little gender discrimination by teachers in class, as there is evidence to suggest that female teachers are neither any better nor worse than their male counterparts with regard to in-class relationships with their students (Odaga and Heneveld 1995) Nevertheless, promotion of female teachers has been commonly recommended as a strategy to encourage girls' education, assuming that apart from providing positive role models to young girls, particularly in rural areas, parents are put at ease about their daughters' safety by the presence of female teachers. However, the findings of Casely-Hayford (2008) in Ghana suggest how the experiences of female teachers in isolated patriarchal rural communities can be very difficult and may not offer positive images to young women at all. Odaga and Heneveld (1995) found some cases where the analyses suggested no difference in low expectations of female students between female or male teachers, and little evidence of the positive impact of female teachers on female students' performance. Similarly, as pointed out by Arnot and Fennell (2008a), it is not evident, but instead, rather unclear, whether the sex of the teacher makes a difference to the educational outcomes and the impact of male and female achievement patterns (ibid. 520). On the other hand, a study from Uganda indicated that the largest gender gaps in enrolment exist in poor regions where the percentage of female teachers is low (Odaga and Heneveld 1995). Obviously, the gender aspect with regard to teachers and teaching is far from clear or univocal.

Female students' educational outcomes, particularly at the post-primary level, are also influenced by the type of educational institution, referring earlier to the public–private debate, here to single-sex or co-educational schools. The consensus in the literature is that girls in single-sex schools tend to perform better in national examinations than those in co-educational schools, particularly in science subjects and mathematics (Odaga and Heneveld 1995). For example, a country case from Kenya evidenced that the smaller the female population in a school, the larger the number of female dropouts; this is suggestive of the importance of peers. In a country case from Malawi, in co-educational schools, female students were in a minority and they also tended to be younger than their male colleagues. They faced harassment, teasing and ridicule from boys, on the one hand, for not being intelligent, but on the other, for being too intelligent, that is, 'unfeminine'. A study from Cameroon verified that female students in good quality co-educational schools performed academically better than males, but that they also generally came from better socio-economic backgrounds (ibid.; cf. Colclough, Rose and Tembon 2000). Similar kinds of patterns characterise also the Tanzanian experiences. Dividing students into separate schools is probably not a sustainable strategy to tackle the issue of safe and conducive learning environments, because at some point, the female and male students will study and work together, as commented on by one of the participants in this research. At the same time, laying the grounds for female students' self-confidence and resilience, and on the basis of educational progression and learning results, single-sex secondary schools have proven benefits, evidenced also in the stories of the participants in this research, as we will hear later. Consequently, different alternatives are on the agenda of education sector reform in Tanzania, and the government is opting, for instance, for gender-streaming as an alternative to single-sex schools, showing promising results in terms of improvements in performance for both sexes, as identified already by Odaga and Heneveld (1995, 28–30).

Furthermore, the government is committed to building single-sex hostels for students studying in co-educational institutions to improve the availability of gender-segregated sanitation facilities, for example, and to provide guidance and counselling to enhance overall gender awareness, among other things, within schools.

The type of school and distance to school are interconnected with safety issues and sexual harassment, mentioned above. The question of the 'distance' is considered to be two-dimensional: one aspect concerns the distance and the energy children have to expend over the distance, often with an empty stomach, while the other relates to (sexual) harassment. As pointed out by Brock and Cammish (1997), the children in the developing world tend to have extremely long school days due to the very long distance to schools, meaning that it is quite common to leave home around 5.30 a.m. and return at 5.30 p.m. This is a barrier to education and schooling that the government of Tanzania is partially tackling with the ambitious school construction work by literally bringing schools into every village. The government has also paid attention to the issue of food by introducing different schemes to provide school meals (free, partially subsidised by the school/government, paid for by the parents); however, the majority of children still seem to spend their school days without any food.

The second concern has been discussed in path-breaking research on 'Sexual Harassment and Violence against Female Students in Schools and Universities in Africa' by Hallam (1994), suggesting that there is a pandemic of sexual violence in educational institutions in Africa. The study conducted by Mbilinyi and Mbughuni (1991) in Tanzania indicated that harassment of schoolgirls and college and university women by fellow students and teachers is a very real concern and requires more investigation. Apart from sexual harassment, and even physical force and rape, Odaga and Heneveld (1995) found evidence of verbal abuse, teasing and threatening in the various studies that they reviewed. Apart from harassment from male students, teachers were found to prey on females, threatening to fail them, or publicly humiliate them to prod them into sexual liaisons (rewarding students who 'cooperate' with good grades and tuition waivers); furthermore, there is evidence that older men abuse young schoolgirls outside school hours. Trading sex for money to cover the schooling costs and being harassed and humiliated by teachers were also exemplified in the research participants' narratives. Similar kinds of notions and concerns are recognised and discussed in studies by Leach in Zimbabwe, Malawi and Ghana (Leach 2003a; see also Leach 2006; Leach and Mitchell 2006), Morley (2011) in Ghana and Tanzania, in Tanzania by Stambach (2000), Unterhalter in South Africa (2003b) and a growing number of other individual researchers and agencies.[1]Dunne, Humphreys and Leach (2006), for example, analyse both the implicit gender violence that relates to everyday structures and practices at schools, and explicit gender violence, which in turn relates to more overtly sexualised encounters. Clearly, gender violence in schools, in its culture, structures and processes, is an extremely delicate area of concern to investigate, but needs to be redressed by educational authorities, as unconditionally concluded by Odaga and Heneveld (1995). They argue that the accounts of harassment and violence indicate that female students face an extremely hostile and uncomfortable learning environment, undermining the wide range of efforts to increase female access to and achievement in education. Additionally, acts of sexual harassment and violence in schools and universities reflect society's negative views of women. Finally, these acts have a profound, devastating and far-reaching effect on the girls, not only on their educational progression, but also on their lives as a whole (ibid.)

Among evident and concrete consequences of sexual conduct are (unexpected and unwanted) pregnancies. As we learned earlier, in primary schools, the main reasons for dropout in Tanzania were truancy (67 per cent), followed by death (5.6 per cent) and pregnancy (5.5 per cent) (URT 2009a) As pointed out by Odaga and Heneveld (1995), pregnancy becomes a major factor in school dropouts as the female students become adolescents, and as evidenced in Tanzania, pregnancy emerges as one of the major causes of female students leaving school. Rightly, fear of pregnancy is a reason why parents remove their daughters from school as they approach or reach puberty, as found by Serpell (1993) when examining schooling and life courses in Africa (see also Stambach 2000). All of the women that I interviewed for this study had known some students in their primary schools who became pregnant and dropped out, and none of these girls had returned to school. In addition, and referring to the 'far-reaching effect' mentioned above, it needs to be noted that during the last years of primary school, the female students are only between 11 and 12 years of age. Partially for this reason, at the level of secondary education, female students' sexual behaviour is usually strictly observed, as we will hear also from the research participants' narratives. As stated by Odaga and Heneveld (1995), this reflects the tendency to blame the girls and young women, without querying the responsibility of the men (whether they are boys or, as is not uncommon, the male teachers at the schools). Thus, from an educational point of view, pregnancy results in expulsion and, consequently, the end of female students' schooling; from a health perspective, the implications of teenage pregnancy include a very high risk of death (including illegal abortions, as witnessed by one of the participants in this research) and illness (including sexually transmitted diseases).

Odaga and Heneveld (1995) also found that female students' own expectations and motivation, reflecting the low expectations at schools and in society at large, when intertwined with the characteristically high levels of repetition, dropout rates and failure, are factors that influence female education and schooling. Hence, to conclude this section concerning the school environment, what is *the idea of Tanzanian woman* that the female students and the society at large have, embedding also the idea of education? It was mentioned briefly above how powerful the images that the female students face in school are in terms of shaping their self-perceptions, including their educational aspirations. In analysing the meaning of schooling in rural Zambia, Serpell, for example, (1993, in Odaga and Heneveld 1995, 41) found the following female students' views of themselves: 'girls in general, and I in particular, do not have the intellectual ability to cope with the curriculum; the most important challenge at this stage in my life is to get married and start a family, and further schooling will contribute little or nothing to my attainment in those goals: indeed, it may even impede it'. Thus, it is evident that the socio-cultural factors, which will be examined in detail in the next section, and the socialisation processes affect female students' self-image, performance and attainment in school and their later career aspirations. Academically, as presented earlier, female students are characterised as lazy and uninterested in school (by the teachers, parents and the students themselves), and because of the lack of interest and poor performance, there is no reason to invest in female education. All in all, the reasons and reasoning to *value* education, or in contrast, to comprehend schooling as a waste of resources, are complexly intertwined with political and institutional factors, school-related factors and, particularly where the girls and the women of the households are concerned, socio-economic and socio-cultural factors. Next, the focus is on understanding the complexities embedded in the latter factors, that is, the social and familial levels of the enabling environment.

Demand for Female Education in Families and Communities

Amongst the most critical findings of the literature review on women's education in sub-Saharan Africa by Hyde (1989a) and Odaga and Heneveld (1995), country case studies focusing on factors affecting female *participation* by Brock and Cammish (1997), and on *attendance and achievements* in developing countries by Hyde (1989b) are, first, that the socio-cultural and socio-economic factors that constrain girls' and women's education and schooling at the household and community level are closely interwoven, and second, that their effects on girls' education are robust and far-reaching, concerning not only access and enrolment, but affecting the performance and persistence of those who remain in school. Third, they concluded that the knowledge and understanding of how these factors govern household decision-making is the key to formulating strategies to address the low societal demand for female education (see also Stromquist 1989).

In Tanzania, the socio-economic and socio-cultural factors have been included in policy-level discussions for over 40 years: they were first raised soon after the country gained independence (see e.g. Buchert 1994); yet recent studies emphasise how similar issues still exist (Colclough et al. 2003). Our observation and concluding remarks in the pre-study of this research (discussed in Chapter 4) are that the current education sector reform and programmes in Tanzania acknowledge the importance of the socio-economic and socio-cultural factors; however, the policy-level initiatives are directed mostly at the school environment. The environment that enables or prevents girls' and women's education and schooling reaches, however, far beyond the education system and schooling as such, and cannot be overlooked if the gendered educational targets and goals are to be reached (Okkolin, Lehtomäki and Bhalalusesa 2010; Colclough, Rose and Tembon 2000).

The social and familial levels of the enabling environment include factors such as poverty, level of parental education, social class, region of residence, familial investment behaviour (reluctance to invest in education; low expectations of the economic value of schooling for girls), the economic value of girls (needs for girls' input in income generation and household activities and care for siblings), familial perceptions of the irrelevance of schooling for girls, and cultural beliefs, and values and norms including views of female life course etc., all affecting girls' and women's education and schooling. The literature on education, gender and development is packed with studies and reports discussing the interlinkages between education, gender and *poverty* (e.g. Chege and Arnot 2012; Colclough, Rose and Tembon 2000; Leach 2003a; Lugg, Morley and Leach 2007; UNICEF 2012a; Vavrus 2002a; Wedgwood 2005). As discussed in relation to the school environment, the literature demonstrates that the lack of resources to cover the direct and opportunity costs of educating girls is a major constraint on girls' education: in SSA the most important reasons for not being enrolled, but also for non-attendance, were the direct costs of schooling, such as fees, uniforms and equipment; from the gender perspective, it was easier for boys to find work and contribute to or cover the costs, whereas girls had more domestic responsibilities and therefore their education was more costly.

Odaga and Heneveld (1995) have pinpointed that we should treat cautiously the studies arguing that the poor performance of children, particularly females, is related to the greater demands on their time for household chores only, because the amount of time they spend on chores, such as taking care of siblings, fetching firewood and water, cooking food, working on the farms, taking care of the cattle, and other productive activities such as marketing/petty trading (to raise money for school costs and/or ensure household survival) simply reduces the time and energy that is available for their schooling (see also Colclough et al. 2000, 2003). They (ibid.) highlight that a much more complex (gendered) process is involved. As pointed out by Semali and Mehta (2012, 228), the number of children participating in economic and housekeeping activities in Tanzania is staggering: in 2006, nearly 40 per cent of children aged 5–17 worked in economic activities and 48 per cent were engaged in housekeeping activities; the participation rate is highest in rural areas (45 per cent), in comparison to 6.5 per cent in Dar es Salaam. They also demonstrate how the participation rate in economic activities increases with age; they note that for housekeeping activities the participation rate peaks at 10–14 and then drops for those who are 15–17 years old (also Colclough, Rose and Tembon 2000). This is not to claim that there is a causal relation between household chores and (low) attendance and (poor) attainment, as we will also learn from the research participants' narratives, but it is plausible to claim that the time allocation has a direct bearing on them and makes a difference from the perspective of educational well-being. It also makes a difference from the gender perspective, if only the girls are given the responsibility of undertaking a vast amount of the household chores and other familial duties. Similarly, explaining and reducing social and familial practices to poverty, although evidently underlying household decision-making processes, is a plain oversimplification of far more complex and intersecting phenomena (e.g. Colclough, Rose and Tembon 2000; Fennell and Arnot 2009; Vavrus 2002a, 2002b; Wedgwood 2005) For instance, the demand for female labour at home is a factor defined socio-culturally rather than economically, and traditions and social roles are more binding in rural than in urban areas; indeed, demand for female education appears to be lowest in rural and marginal areas where poverty is most endemic, and where opportunities for income generation are limited (Arnot et al. 2012a; Caplan 1995; Stambach 2000; Unterhalter 2012; Vuorela 1995; cf. Helgesson 2006; Moyer 2003; Shabaya and Konadu-Agyemang 2004; Hansen 2005)

The connection between poverty and education has been discussed by Wedgwood (2005) and Lewin (2009), and the gendered outcomes of poverty in education by Colclough and colleagues (Colclough, Rose and Tembon 2000), Swainson (2000), Vavrus (2002a), Shabaya and Konadu-Agyemang (2004) and Morley, Leach and Lugg (2009), to name just a few. In brief, what Wedgwood says is that poverty is commonly given as an explanation for not being educated, and not *vice versa*. Referring to a Participatory Poverty Analysis in Tanzania, in which a sample of poor rural people gave their views on what makes them poor, the most common statement was that 'I am uneducated because I am poor' rather than 'I am poor because I am uneducated' (Wedgwood 2005, 11). Lewin's (2009) findings suggest that school participation remains very uneven when analysed in relation to household income, especially at the secondary level of education, and being able to progress further is most powerfully determined by wealth. Colclough, Rose and Tembon (2000, 5) claim that presuming (as is commonly done) that under-enrolment and the difference in access to and achievement in education for female and male students is primarily caused by poverty means that the gender gap could be expected to disappear as development progresses. They continue by stating that, as the level of poverty decreases, the tendency is that school enrolments increase; however, the progress towards gender equality is slow. Furthermore, if gender differences are caused by a set of factors other than poverty, both quantitative and qualitative inequities in school outcomes are likely to remain entrenched (as commonly happens).

Similarly Swainson (2000) has pointed out how the common assumption that increasing the overall supply of primary education (by means of high public expenditure) will eliminate gender gaps is seriously flawed, because despite increased female participation in education, for example in Tanzania, the deep-rooted problems remain with regard to the poor educational outcomes accompanied by the low self-image of girls. Vavrus (2002a), however, argues strongly for the compounding effects of the opportunity costs, and places the emphasis on research findings that do not embrace gendered and discriminatory ideas towards female education and schooling. Shabaya and Konadu-Agyemang (2004) draw together the various spatial and socio-economic dimensions, most importantly, the costs of schooling and the priority given to the idea of women as future wives and mothers, as the main causes of the gender gap in Ghana, Zimbabwe and Kenya. Similarly, Morley and her colleagues (2009, 62–63) found out that the opportunity structures for education in Ghana and Tanzania reflect poverty *and* wider social inequalities, and higher education in the two countries is still highly selective; hence, their findings are suggestive of conversion and consolidation of economic wealth into educational advantage (see also Lewin 2009 and Stromquist 1989). Morley, Leach and Lugg (ibid. 57) also reminds us how poverty is increasingly perceived as a capability deprivation, that is, a constraint on people's opportunities to pursue and achieve various 'beings and doings' that they have a reason to value – as will be discussed later.

Overall, as summarised by Odaga and Heneveld (1995), girls are an important source of income for their families, and the need for additional household income often takes priority over education. Yet the cultural and economic factors are intersecting, and the relationship between poverty, social class and gender depends on the specific contexts. Generally speaking, the socio-cultural expectations for the future role of girls, that is, marriage and family, as was previously noted, in addition to the high social status attached to marriage and motherhood and the contrasting pervasive gender ideologies at the community level, which often favour males over females (such as, boys are more intelligent, they perform better, and are, in consequence, a much better educational investment), combined with some communities holding negative views of educated girls, believing that education may push them towards prostitution, that they are unfaithful, difficult to control, too independent and demanding, which, in turn, diminishes their marriageability (and bride price). All this depresses the demand for female education and promotes differentials in educational opportunities and outcomes (ibid.)

The views about what educated girls and boys ought to learn and become in the Kilimanjaro region and Chagga society in Tanzania have been interestingly explored by Stambach (2000), where she came across ideas that resonate with and amplify the portrayals discussed above, but where she also uncovered very different kinds of ideas: to some, schooling signified the demise of what they saw and valued as traditional culture; to others, education and schooling were a sign of social progress (and a ladder to higher social and economic levels); for some, schooling meant disrupting appropriate gendered relations, whereas for others, schooling was viewed as a desirable institution through which 'modern families' could be created, nurtured and reproduced (ibid. 3). Positive ideas concerning girls' and women's education and schooling were also identified in Odaga and Heneveld's (1995) review, and they pointed out how men may actually prefer girls who have gone to school, and how some parents, particularly mothers, prefer investing in girls' education for the very same reasons that are often given for not investing in them: a more secure family and old-age support, future mothers who would require money to look after their families etc. All in all, the information campaigns, the encouraging examples and the overall awareness regarding the value of education in general, and the importance of educating girls to achieve progress in particular, have all, according to their review (ibid.) impacted positively on gender equality in education. Similar kinds of examples of parental support and encouragement to educate daughters have been found by Morley, Leach and Lugg (2009) when identifying the 'opportunity structures' in Ghana and Tanzania, and by Warrington (2013) when examining the different roles of fathers and mothers in Uganda. The importance of familial and guardianships support, and the exposure and environmental factors, in addition to personal resilience, were summarised in Kakenya's (2011) findings as the most important factors that contributed to Maasai women in Kenya succeeding in education (see also Warrington and Kiragu 2012; cf. Omwami 2011). Vavrus (2002a), in turn, refers to the findings of various demographic studies of girls' education in which the mother's level of education is correlated with the level of education of her children, and in which the data also suggest the impact of the father's educational background on the educational attainment of adolescent girls; her examination found that in the Kilimanjaro region, secondary school girls tended to have fathers and mothers with post-primary education. The findings from Ethiopia of Rose and Al-samarrai (2001) also proved that the initial demand for girls' education, in particular, was positively influenced by the parental attitudes acquired through their own education, albeit that the issues of the nutritional status of the child, economic status, age and puberty/marriage all seemed to impact on the attendance and completion of education. All of these ideas and notions are also identifiable in my study, as we will hear later from the research participants' narratives.

As argued by Odaga and Heneveld (1995), and as discussed already, social-cultural customs and beliefs influence parental decisions to enrol girls in school and in their decisions to withdraw them from school; they impact on female students' own decisions to drop out, their academic performance and grade-level attainment, and thus on their position in the labour market and on their employment prospects in general (cf. Bendera 1999). In general, female students in the developing countries tend to aspire to traditional female occupations, regardless of the place of residence, and in rural areas, poor female students have an even more limited range of career aspirations, with the professions of teaching and nursing being the most popular ones (Odaga and Heneveld 1995). By chance, one of the women participating in this study spoke of how she had primarily wanted to become a nurse (because of the uniform and the appearance of nurses), but ended up in the teaching profession. Yet the fact that society, parents, teachers and the students themselves have a certain kind of *idea of women* (that is, by definition, low expectations and assumptions regarding the quantity and quality of female education) reinforces and supports their low academic performance and high dropout rates; the lack of parental interest and the lack of interest by students themselves may also lead to truancy, reinforcing low levels of academic progression, and contributing to the idea of educational wastage at the primary and secondary levels of education from the parents' perspective. Furthermore, parents may lack control over their children's school attendance, who themselves may consider schooling as boring, and the children may view making money as much more interesting On top of that, the generally positive perception of education as a means to a better life is impaired by the reality of uncertain educational outcomes, student's experiences with repetition and failure, and dire poverty, where the immediate daily struggle outweighs the longer-term potential benefits of education (ibid.) Therefore, education and schooling is a highly contradictory project for parents and students alike, and there seems to be a rather 'vicious circle' characterising the education and schooling of girls and women in sub-Saharan Africa.

In addition to poverty, the socio-economic status and parental education level, rural/urban residence, the economic value of girls and parental and familial investment behaviour (expectations of the value of education), ideas of the irrelevance of schooling, the value placed on marriage and motherhood, future employment prospects etc. discussed previously, Table A.1 lists other factors that influence female education, such as initiation ceremonies, religious beliefs and HIV/Aids. Odaga and Heneveld (1995) remind us how initiation ceremonies are important in some sub-Saharan African countries, including Tanzania (referring to a study carried out by the Tanzania Gender Networking Programme, a famous and well-recognised activist organisation), where 'training' was the most common cause of dropout for both boys and girls, suggesting a very low value being placed on formal education. They also pay attention to the psychological impact of the rites, in terms of being defined as a child or an adult, how the children are to be treated in school, and how the children behave towards their teachers and peers. On the other hand, they note how the rites, particularly circumcision, may have drastic mental and physical consequences. Moreover, from a more practical point of view, as with the issue of gender-segregated sanitary facilities, the initiation ceremonies may cause problems for female students' attendance and performance, and may even lead to dropout due to clashes between ceremonies and the school calendar. The issue of religion is usually associated with low female participation in schools, but in the Tanzanian context, the importance of religion and missionaries is often given as an explanation and a reason for the educational success of the Kilimanjaro region, for instance, as has been briefly referred to above and as is also mentioned later. The problems caused by the HIV/Aids pandemic in the African continent for individuals, individual families and for the countries are immense and are taken explicitly into consideration, for example, in the Millennium Declaration (Goal 6: Combat HIV/AIDS, malaria and other diseases). Furthermore, the list of issues that influence female education seems to be endless, and includes the (lack of) opportunities for further education, (lack of) role models, pregnancies, over-age enrolment and very practical issues such as hunger and illness (i.e. who eats and when in the family) etc., which all have the tendency to have a strong negative bearing on female students' formal educational opportunities (Odaga and Heneveld 1995) Consequently, as was pointed out at the beginning of this section, the most challenging factor in tackling gender inequality in education is its characteristically intertwining and intersecting nature.

As concluded by Odaga and Heneveld (ibid.), over 20 years ago, the complexity of household and familial perceptions of female education and familial arrangements, and that complexity's influence in shaping the low investments in formal education for girls is beginning to be given due attention in policy and research. More recently, Stambach (2000), among others, has noted how the schools are often pivotal social institutions around which the configuration of society as a whole is imagined, contested and transformed; and how schools also provide one of the clearest institutions for observing debates on culture, generation, gender and history. Evidently, both schools and social and familial levels of the enabling environments embed a myriad of cultural institutions and practices that limit the potential for formal female education, but there are also factors that may widen the opportunities for pursuing gender equality in education, and this perspective warrants further research, and is the approach employed in my study.

Assessing Gender Equality and Equity in Education

Against the broader discussion and research on education, gender and development presented above, how then are the achievements and challenges regarding gender equality and equity assessed in the mainstream development narratives? I will begin by introducing 'women' and 'gender' into the development discourse and agenda. Then, to broaden the focus and scope of the mainstream methodology of assessment, I introduce briefly the key ideas, elements and concepts of the capability approach (to posit my study in the framework of 'equality of capabilities', to which I return in Chapter 7).

Gross and Net Enrolment Rates (GER and NER) and gender parity are the most widely used indicators that the national bodies and international agencies apply to monitor and evaluate educational advancement.[2] These indicators are, however, static 'snapshots' and hence problematic, because neither the GER or NER can monitor, for instance, how regularly the children actually attend school, nor do they capture the aspects of the quality of teaching and learning, which are critical for educational attainment.

In the national statistical data in Tanzania, some educational indicators, such as enrolment, performance and completion rates, were monitored and presented according to gender, although unsystematically. However, as noted by Unterhalter and Brighouse (2007) and Subrahmanian (2005) *gender parity* in 'access to' and 'participation in' education is neither synonymous with nor an index of *gender equality*, nor does it capture the complexity of intersecting and relational dimensions of *gender inequality*. For example, the Education and Training Policy in Tanzania (URT 1995), MDGs and EFA entail both gender parity and gender equality goals, characterised as quantitative and qualitative goals respectively. Presumably, as remarked on by Subrahmanian (2005), both kinds of assessment are needed to monitor and evaluate progress towards both numerical and qualitative goals. However, as she points out, it is somewhat disconcerting that the Dakar Framework of Action, for instance, does not actually define the concept of gender equality, hence making the measurement of progress towards EFA goals hard, if not impossible, and a subject of contestation and dispute (Subrahmanian 2005; see also the post-MDGs arguments made by Fukuda-Parr 2012 and Unterhalter 2014). What Subrahmanian proposes is that gender equality in education should be understood as the rights to, within and through education (compare Vaughan 2007). At the core of this argument, for Subrahmanian (2005), is the recognition of *substantive equality* (e.g. Kabeer 1994; Kabeer and Subrahmanian 1996), which is also the essence of the capability approach (e.g. Sen 1992, 1993, 1999; Robeyns 2005, 2006).

It is evident that different understandings of gender, equality and education generate different approaches with which to pursue gender equality in education; different ways in which to assess development and change; and, finally, different portrayals of gender equality in education *par excellence*. These understandings have been examined by Unterhalter (e.g. 2004, 2007). Next, I will present her idea of the contrasting approaches relating to gender equality in the educational sphere, since this division partially justifies and posits the premises and methodology adapted in my study. According to Unterhalter (2007), there are at least three ways to understand gender equality. First, gender can be understood as a descriptive characteristic of a person based ontologically on biology, and, as she states, 'meaning no more than having short hair or a birthday in September' (ibid.). Within this approach, equality in education is connected first and foremost to distribution, such as the availability of schools (for all) and an equal number of years of schooling. This understanding dominates and guides the ways in which the governments collect and present statistical data and international organisations such as UNESCO and the OECD make comparisons (Unterhalter 2004; Okkolin and Lehtomäki 2005). This is also the mainstream approach to assessing progression in Tanzania, by monitoring the rates and parity in 'access to' and 'participation in' education. On the basis of this first understanding and approach, where gender equality signifies an equal number of girls and boys enrolled in school, attending classes, performing and completing examinations, Tanzania is doing well in primary education. According to Subrahmanian (2005, 397), this is not, however, gender equality, but gender parity, offering only the 'first-stage' measure of progress towards gender equality. Unterhalter (2007) has termed this approach *distributive equality* (the resourcist approach in her earlier work), having its roots in and connections with significantly the 'women in development' approach that emerged in the 1970s.[3]

The main targets for the first wave of feminism in development and 'women in development' (WID) advocates were to make women a recognised constituency, a visible and distinctive category in development policy and discourse, in response to the denial of the voice and agency of women (see e.g. Fennell and Arnot 2009; Kabeer 1994). Quite soon, however, it became clear that this recognition was rather categorical and symbolic, and that it was unsupported by material resources or political commitment, and this paved the way for the shift from 'women' to 'gender relations' as the focus of analysis (Kabeer 1994, xii) The implication of relying categorically on 'women' only, was that the 'problem' too, for instance, of achieving gender equality in education, like the solution, concerned women only, and put aside the very essence of the 'problem', that is, its relational and hierarchical nature. For this reason, the gender and development (GAD) approach, having a more structuralist perspective with which to re-examine similar kinds of 'problems' that the WID advocated had sought to address, took root in the development discourse and agenda. As remarked on by Kabeer (1994, xiii), 'just as a class analysis can be used to understand and address the problems of the poor, so too a gender analysis can be used to understand and address the problems of women's subordination'. Thus, having its roots in and connections with Marxist thinking, but moving well beyond the tradition, the analyses moved towards structured and socially constructed disadvantages that women face relative to men, drawing attention to positions constituted by social relations. The constructionist ontological orientation lays the foundation and premises for this examination, the emphasis being profoundly on human agency, as will be explicated later.

The GAD approach resonates with the second understanding of gender equality in education, defined by Unterhalter (2007) as *equality of conditions*, and deriving from structuralist and quite different post-structuralist approaches (associated ideas deriving from critical theory and post-colonialism; the former are briefly touched upon later, the latter I have decided not to discuss). Evidently, and in contrast to the value-neutral first meaning with regard to 'having a birthday in September', these approaches are packed with structural inequalities and discriminatory processes, practices and meanings, on which basis gender is ontologically constituted and constructed. Hence, gender equality means the removal of structural barriers and institutionalised discriminatory values, norms, procedures and rules, although, as suggested by Unterhalter 2004, the post-structuralist discourse tends to place more emphasis on multiplicity and the prospect of *difference* rather than of equality, and thus to complex forms of negotiations exercised by the excluded groups, such as women. The key point and methodological challenge here is that gender equality is not just about counting numbers, nor is it about an equal amount of education, equality of outcomes or achievements; rather, it is about creating equal empowerment and enabling conditions for people which, in turn, requires creating an understanding of the complex array of social relations and arrangements embedded in the girls' and women's education and schooling. In other words, within the *equality of conditions* approach, gender resonates with deeply imbued social relationships and positions, while equality refers to equalised conditions in school and education in its broadest sense (interlinked with other areas of social policy); similarly, it refers to 'equal regard', influenced by an appreciation of considerable difference (Unterhalter 2007). As remarked on by Unterhalter (ibid.), the vision of equality here is enormously important as a goal to be aimed at, yet it might be too overwhelming a policy challenge for governments.

Kabeer (1999, in Subrahmanian 2005) has made a statement according to which the achievement of substantive equality requires the recognition of the differences between the capacities of women and men based on biology, *and* in terms of socially constructed disadvantagement. To continue with this idea, Subrahmanian (2005) has proposed that achieving substantive equality depends on two further processes to reveal *how* it has been reached. She (ibid.) suggests focusing on the socially constructed pathway(s) to equality, referring at first to the quality of experience (access to education, participating in it and benefitting from it), presuming that there is equality of treatment and equality of opportunity, and second, to a commitment to non-discrimination. However, assessing gender equality, reaches beyond the aspects of treatment and opportunity, and involves the elements of agency and autonomy. In other words, for Subrahmanian (ibid.), to achieve and to assess substantive equality, and for the process to be meaningful from the perspective of gender, depends profoundly on the equal distribution and availability of resources and the recognition of fundamental freedoms.

What Subrahmanian asserts corresponds with the third meaning of gender equality in Unterhalter's definition, arguing that gender implies both socially constructed constraints on women's *capability sets* and the negotiation processes dependent on women's *agency* and *freedom*. According to her, the third meaning of gender equality, resonating with and utilising the vocabulary of the capability approach, has affinities with the two first meanings, yet is fundamentally distinctive (to which I will return in the next section). The key here is the possibility for change, referring to the attributes of the person, on the one hand, and to social relations, on the other. For instance, girls from different families and broader social and societal contexts might 'do gender' differently at different moments in their own lives, sometimes because of the demands of the structures, but sometimes because of negotiations and contestations; this meaning of gender acknowledges the power of macro-level social relations, yet, analogously to gendered micro-identities, comprehends the gendered educational hierarchies and institutions to be challengeable and changeable. (Unterhalter 2007). Unterhalter relates the third meaning of gender equality to the concept of representational justice, referring particularly to the work of Nancy Fraser, who locates the dimensions of process and difference at the core of equality assessment by comparing i) whether women participate in terms of parity with men in decision-making and ii) whether the languages and routes to power they have are equal. Thus, when framed within the discussion and assessment of gender equality in education, the focus is on the micro-level pathways and processes, and on the macro-level commitments to non-discriminatory policies in education, acknowledging that different groups of people, such as female students, might need different kinds of initiatives to advance (ibid. 91; cf. Subrahmanian 2005).

Obviously, this understanding of the equalities in education overlaps with the second definition, and elsewhere, too, the conceptual boundaries and the approaches *per se* cross over. *In practise*, gender inequalities cannot be addressed by any single approach to gender or equality, but all three initiatives need to 'overlap' and complement each other to enhance policy and practice (Unterhalter 2007, 104). Yet in Tanzania, , for instance, the focus is notably on the realisation of distributive equality; some resonance and linkages with the equality of conditions approach can be found but very little recognition is given to the equality of capabilities approach. As stated by Moser (2007, 10), deciding what to assess is a political process; yet in practice, assessment depends on objectives and goals; it also depends on the identification of the changes that are required to achieve these goals; and finally, it depends on the decisions regarding aspects and indicators that enable the measurement of progress towards the desired changes. Consequently, in practice, referring to Moser's notions and the previous discussion on different kinds of understandings of gender equality in education, it is reasonably clear that it is much easier to assess development when it is understood as a change in numbers and amounts, and, in contrast, it is more complicated if the change is targeted towards the less explicitly defined goal of equality. If, for example, the perspective on gender equality in education is broadened, and defined as concerning issues such as alternatives, choices and decision-making processes, in other words, opportunities for and constraints on educational *well-being* and *agency* (including a social and societal ethos towards girls' and women's education and schooling), as in my study, then there are areas under consideration that are not reachable only on the basis of the information-base characteristics of the 'distributive equality' and 'equality of conditions' approaches alone. Consequently, there is a need to broaden the information base accordingly. Before introducing a detailed explanation of how the 'equality of capabilities' approach is analytically applied in my study, a review of education policies, strategies and instruments to promote gender equality and equity, *and* a situational analysis regarding achievements and challenges of the recent education sector reforms in the Tanzanian context are presented.

Notes

1 See e.g. the Study on Violence against Children by the Centre for International Education at the University of Sussex: http://www.sussex.ac.uk/cie/projects/comp leted/genderviolence
2 The gross enrolment rate indicates the number of children attending school (in a census) regardless of their age, for which reason the rate is often above 100 per cent due to a large number of under-aged and over-aged children. The net enrolment rate in turn, is adjusted to the official school age, being for example 7–13 for primary education in Tanzania.

3 The emergence and evolvement of gendered thinking and discourse in the development agenda is discussed comprehensively by Kabeer (1994), Kabeer and Subrahmanian (1996) and Moser (1993). See also the comparison between UNESCO and the World Bank understandings and approaches concerning girls' and women's education by Peppin Vaughan (2010).

Bibliography

Ampiah, J.G., and C. Adu-Yeboah. 2009. Mapping the Incidence of School Dropouts: a Case Study of Communities in Northern Ghana. *Comparative Education* 45, no. 2: 219–232.

Arnot, M., and S. Fennell. 2008a. Gendered Education and National Development: Critical Perspectives and New Research. *Compare: A Journal of Comparative and International Education* 38, no. 5: 515–523.

Arnot, M., R. Jeffery, L. Casely-Hayford, and C. Noronha. 2012a. Schooling and Domestic Transitions: Shifting Gender Relations and Female Agency in Rural Ghana and India. *Comparative Education* 48, no. 2: 181–194.

Barrett, A. 2007. Beyond the Polarisation of Pedagogy: Models of Classroom Practice in Tanzanian Primary Schools. *Comparative Education* 43: 273–294.

Bendera, S. 1999. Promoting Education for girls in Tanzania. In *Gender, Education and Development. Beyond Access to Empowerment*, ed. C. Heward and S. Bunwaree, 117–132. London: Zed Books.

Bennell, P., and F. Mukyanuzi. 2005. *Is There a Teacher Motivation Crisis in Tanzania?* Brighton: Knowledge and Skills for Development.

Brock, C., and N.K. Cammish. 1997. *Factors Affecting Female Participation in Seven Developing Countries*. Vol. 9 of *Education Papers*. 2nd ed. London, UK: DFID.

Buchert, L. 1994. *Education in the Development of Tanzania 1919–1990*. Athens, OH: Ohio University Press.

Caplan, P. 1995 'Children are Our Wealth and We Want Them': A Difficult Pregnancy on Northern Mafia Island, Tanzania. In *Women Wielding the Hoe: Lessons from rural Africa for Feminist Theory and Development Practice*, ed. D.F. Bryceson, 131–149. Oxford: Berg Publishers.

Casely-Hayford. L. 2008. Gendered Experiences of Teaching in Poor Rural Areas of Ghana. In *Gender Education and Equality in a Global Context: Conceptual Frameworks and Policy Perspectives*, ed. S. Fennell and M. Arnot, 146–162. London: Routledge.

Chege, F.N., and M. Arnot. 2012. The Gender–Education–Poverty Nexus: Kenyan Youth's Perspective on Being Young, Gendered and Poor. *Comparative Education* 48, no. 2: 195–209.

Colclough, C., S. Al-Samarrai, P. Rose, and M. Tembon. 2003. *Achieving Schooling for All in Africa: Costs, Commitment and Gender*. Ashbourne, UK: Ashgate.

Colclough, C., P. Rose, and M. Tembon. 2000. Gender Inequalities in Primary Schooling. The Roles of Poverty and Adverse Cultural Practice. *International Journal of Educational Development* 20: 5–27.

Dunne, M., S. Humphreys, and F. Leach. 2006. Gender Violence in Schools in the Developing World. *Gender and Education* 18, no. 1: 75–98.

FAWE. 2001. *In Search of an Ideal School for Girls*. Nairobi, Kenya: FAWE.

FAWE. 2003. *The ABC of Gender Responsive Education Policies: Guidelines for Developing Education for All Actions Plans*. Nairobi, Kenya: FAWE Centre of Excellence. http://www.fawe.org/Files/fawe_best_practices_-_centres_of_excellence.pdf.

FAWE. 2006. *Experiences in Creating a Conducive Environment for Girls in Schools.* Nairobi, Kenya: FAWE.

Fennell, S., and M. Arnot. 2009. Decentring Hegemonic Gender Theory: The Implications for Educational Research. RECOUP Working Paper No. 21. London, UK: DFID.

Fukuda-Parr, S. 2012. *Recapturing the Narrative of International Development.* Geneva: UNRISID.

HakiElimu. 2005. *Three Years of PEDP Implementation: Key Findings from Government Reviews.* Dar es Salaam, Tanzania: HakiElimu.

HakiElimu. 2010. *Education in Reverse: Is PEDP II Undoing the Progress of PEDP I? Brief No.10.1E.* Dar es Salaam, Tanzania: HakiElimu.

Hallam, S. 1994. Crimes without Punishment: Sexual Harassment and Violence against Female Students in Schools and Universities in Africa. Discussion Paper No. 4. London: Africa Rights.

Hansen, T.K. 2005. Getting Stuck in the Compound: Some Odds Against Social Adulthood in Lusaka, Zambia. *Africa Today* 51, no. 4: 3–16.

Hardman, F., J. Abd-Kadir, and A. Tibuhinda. 2012. Reforming Teacher Education in Tanzania. *International Journal of Educational Development* 32: 826–834.

Helgesson, L. 2006. *Getting Ready for Life: Life Strategies of Town Youth in Mozambique and Tanzania.* PhD diss., Umeå University.

Hyde, K. 1989a. Improving Women's Education in Sub-Saharan Africa: A Review of the Literature. PHREE Background Paper No 89/15. Washington D.C.: World Bank.

Hyde, K. 1989b. *Improving Girls' School Attendance and Achievement in Developing Countries: A Guide to Research Tools.* Washington D.C.: US Agency for International Development.

Kabeer, N. 1994. *Reversed Realities: Gender Hierarchies in Development Thought.* London and New York: Verso.

Kabeer, N., and R. Subrahmanian. 1996. Institutions, Relations and Outcomes: Framework and Tools for Gender-Aware Planning. Discussion Paper No. 357. Brighton, UK: Institute of Development Studies.

Kakenya, E. N. 2011. *Warrior's Spirit: The Stories of Four Women from Kenya's Enduring Tribe.* PhD diss., University of Pittsburgh.

Leach, F. 2003a. Learning to be Violent: the Role of the School in Developing Adolescent Gendered Behaviour. *Compare: A Journal of Comparative and International Education* 33, no. 3: 385–400.

Leach, F. 2006. Researching Gender Violence in Schools: Methodological and Ethical Considerations. *World Development* 34, no. 6: 1129–1147.

Leach, F., and C. Mitchell, eds. 2006. *Combating Gender Violence in and around Schools.* Stoke-on-Trent, UK: Trentham.

Lewin, K.M. 2007. *Improving Access, Equity and Transitions in Education: Creating a Research Agenda.* No. 1 of *CREATE Pathways to Access Series.* Brighton, UK: CREATE.

Lewin, K.M. 2009. Access to Education in Sub-Saharan Africa. Patterns, Problems and Possibilities. *Comparative Education* 45, no. 2: 151–174.

Lugg, R., L. Morley, and F. Leach. 2007. *Widening Participation in Higher Education in Ghana and Tanzania: Developing an Equity Scorecard. Country Profiles for Ghana and Tanzania: Economic, Social and Political Contexts for Widening Participation in Higher Education.* London, UK: DFID. http://r4d.dfid.gov.uk/PDF/Outputs/ESRC_DFID/60335-working_paper_2.pdf.

Lwaitama, A., G. Mtalo, and L. Mboma, eds. 2001. *The Multi-Dimensional Crisis in Education in Tanzania: Debate and Action*. Dar es Salaam, Tanzania: University of Dar es Salaam Convocation.

Mbilinyi, M., and P. Mbughuni, eds. 1991. *Education in Tanzania with a Gender Perspective: Summary Report*. Dar es Salaam and Stockholm: SIDA.

Morley, L. 2011. Sex, Grades and Power in Higher Education in Ghana and Tanzania. *Cambridge Journal of Education* 41, no. 1: 101–115.

Morley, L., F. Leach, and R. Lugg. 2009. Democratising Higher Education in Ghana and Tanzania: Opportunity Structures and Social Inequalities. *International Journal of Educational Development* 29: 56–64.

Moser, A. 2007. *Gender and Indicator: Overview Report*. Brighton, UK: Institute of Development Studies.

Moser, C.O.N. 1993. *Gender Planning and Development: Theory, Practice & Training*. London and New York: Routledge.

Motala, S., V. Dieltiens, and Y. Sayed. 2009. Physical Access to Schooling in South Africa: Mapping Dropout, Repetition and Age-Grade Progression in Two Districts. *Comparative Education* 45, no. 2: 251–263.

Moyer, E. 2003. *In the Shadow of the Sheraton. Imagining Localities in Global Spaces in Dar es Salaam, Tanzania*. PhD diss., University of Amsterdam.

Odaga, A. and W. Heneveld. 1995. Girls and School in Sub-Saharan Africa, From Analysis to Action. World Bank Technical Paper No. 298. Washington D.C.: World Bank.

Okkolin, M.-A., and E. Lehtomäki. 2005. Gender and Disability – Challenges of Education Sector Development in Tanzania. In *University Partnerships for International Development, Finnish Development Knowledge (FFRC-publications)*, ed. O. Hietanen, 204–216. Turku, Finland: Turku School of Economics and Business Administration.

Okkolin, M.-A., E. Lehtomäki, and E. Bhalalusesa. 2010. Successes and Challenges in the Education Sector Development in Tanzania: Focus on Gender and Inclusive Education. *Gender and Education* 22, no. 1: 63–71.

Omwami, E.M. 2011. Relative-Change Theory: Examining the Impact of Patriarchy, Paternalism, and Poverty on the Education of Women in Kenya. *Gender and Education* 23, no. 1: 15–28.

Peppin Vaughan, R. 2010. Girls' and Women's Education within Unesco and World Bank, 1945–2000. *Compare: A Journal of Comparative and International Education* 40, no. 4: 405–423.

Räsänen, R. 2007. International Education as an Ethical Issue. In *The SAGE Handbook of Research in International Education*, ed. M. Hayden, J. Levy, and J. Thompson, 58–69. London: Sage.

Robeyns, I. 2005. The Capability Approach: A Theoretical Survey. *Journal of Human Development* 6, no. 1: 93–114.

Robeyns, I. 2006. The Capability Approach in Practice. *Journal of Political Philosophy* 14, no. 3: 351–376.

Rose, P., and S. Al-samarrai. 2001. Household Constraints on Schooling by Gender: Empirical Evidence from Ethiopia. *Comparative Education Review* 45, no. 1: 1–25.

Semali, L.M., and K. Mehta. 2012. Science Education in Tanzania: Challenges and Policy Responses. *International Journal of Educational Research* 53: 225–239.

Sen, A. 1992. *Inequality Re-examined*. Oxford: Oxford University Press.

Sen, A. 1993. Capability and Well-being. In *The Quality of Life: Studies in Development Economics*, ed. M. Nussbaum and A. Sen, 30–53. Oxford: Oxford University Press.

Sen, A. 1999. *Development as Freedom.* Oxford: Oxford University Press.

Serpell, R. 1993. *The Significance of Schooling: Life Journeys in an African Society.* Cambridge, UK: Cambridge University Press.

Shabaya, J., and K. Konadu-Agyemang. 2004. Unequal Access, Unequal Participation: Some Spatial and Socio-economic Dimensions of the Gender Gap in Education in Africa with Special Reference to Ghana, Zimbabwe and Kenya. *Compare: A Journal of Comparative and International Education* 34, no. 4: 395–424.

Stambach, A. 2000. *Lessons from Mount Kilimanjaro: Schooling, Community, and Gender in East Africa.* New York and London: Routledge.

Stromquist, N. 1989. Determinants of Educational Participation and Achievement of Women in the Third World: A Review of the Evidence and a Theoretical Critique. *Review of Educational Research* 59, no. 2: 143–183.

Subrahmanian, R. 2005. Gender Equality in Education: Definitions and Measurements. *International Journal of Educational Development* 25: 395–407.

Sumra, S. 2005. *The Living and Working Conditions of Teachers in Tanzania: A Research Report.* Dar es Salaam, Tanzania: HakiElimu.

Swainson, N. 2000. Knowledge and Power: The Design and Implementation of Gender Policies in Education in Malawi, Tanzania and Zimbabwe. *International Journal of Educational Development* 20: 49–64.

Tao, S. 2013. Why Are Teachers Absent? Utilising the Capability Approach and Critical Realism to Explain Teacher Performance in Tanzania. *International Journal of Educational Development* 33: 2–14.

Thomas, M.A.M., and A. Rugambwa. 2011. Equity, Power and Capabilities: Constructions of Gender in a Tanzanian Secondary School. *Feminist Formations* 23, no. 3: 153–175.

Towse, P., D. Kent, F. Osaki, and N. Kirua. 2002. Non-graduate Teacher Recruitment and Retention: Some Factors Affecting Teacher Effectiveness in Tanzania. *Teaching and Teacher Education* 18, no. 1: 637–652.

UNICEF. 2012a. *Cities and Children: The Challenge of Urbanisation in Tanzania.* Dar es Salaam, Tanzania: UNICEF.

UNICEF. 2012b. *The State of the World's Children 2012: Children in an Urban World.* New York: UNICEF.

Unterhalter, E. 2003b. The Capabilities Approach and Gendered Education: an Examination of South African Complexities. *Theory and Research in Education* 1, no. 1: 7–22.

Unterhalter, E. 2004. Gender Equality and Education in South Africa: Measurements, Scores and Strategies. Paper presented at the British Council/HSRCConference on Gender Equity in Education, May, in Cape Town, South Africa.

Unterhalter, E. 2007. Gender Equality, Education, and the Capability Approach. In *Amartya Sen's Capability Approach and Social Justice in Education,* ed. M. Walker and E. Unterhalter, 87–107. New York: Palgrave Macmillan.

Unterhalter, E. 2012. Inequalities, Capabilities and Poverty in four African Countries: Girls' Voice, Schooling, and Strategies for Institutional Change. *Cambridge Journal of Education* 42: 307–325.

Unterhalter, E. 2014. Walking Backwards into the Future: a comparative perspective on education and a post-2015 framework. *Compare,* 44 no. 6: 852–873.

Unterhalter, E., and H. Brighouse. 2007. Distribution of What for Social Justice in Education. In *Amartya Sen's Capability Approach and Social Justice in Education,* ed. M. Walker and E. Unterhalter, 68–86. New York: Palgrave Macmillan.

Unterhalter, E., A. North, M. Arnot, J. Parkes, C. Lloyd, L. Moletsane, E. Murphy-Graham, and M. Saito. 2014. *Interventions to enhance girls' education and gender equality. Education Rigorous Literature Review.* London: Department for International Development.

URT (United Republic of Tanzania). 1995. *Education and Training Policy (ETP).* Dar es Salaam, Tanzania: Ministry of Education and Culture.

URT. 2009a. *Basic Education Statistics in Tanzania (BEST) 2003–2007 National Data.* Dar es Salaam, Tanzania: Ministry of Education and Culture.

Vaughan, R. 2007. Measuring Capabilities. In *Amartya Sen's Capability Approach and Social Justice in Education*, ed. M. Walker and E. Unterhalter, 109–130. New York: Palgrave Macmillan.

Vavrus, F. 2002a. Making Distinctions: Privatisation and the (Un)educated Girl on Mount Kilimanjaro, Tanzania. *International Journal of Educational Development* 22: 527–547.

Vavrus, F. 2002b. Uncoupling the Articulation Between Girls' Education and Tradition in Tanzania. *Gender and Education* 14, no. 4: 367–389.

Vavrus, F. 2009. The Cultural Politics of Constructivist Pedagogy: Teacher Education Reform in the United Republic of Tanzania. *International Journal of Educational Development* 14, no. 1: 65–73.

Vohua, S., S. Kiragu, and M. Warrington. 2012. *Gender in East Africa: Teaching Against the Odds.* No. 10 of *CCE Report.* Cambridge, UK: Centre for Commonwealth Education.

Vuorela, U. 1995. Truth in Fantasy: Story-telling With and About Women in Msoga Village, Tanzania. In *Women Wielding the Hoe: Lessons from Rural Africa for Feminist Theory and Development Practice*, ed. D. F. Bryceson, 237–253Oxford: Berg Publishers.

Warrington, M. 2013. Challenging the Status Quo: The Enabling Role of Gender Sensitive Fathers, Inspirational Mothers and Surrogate Parents in Uganda. *Educational Review*, 65, no. 4: 402–415.

Warrington, M., and S. Kiragu. 2012. 'It Makes More Sense to Educate a Boy': Girls 'Against the Odds' in Kajiado, Kenya. *International Journal of Educational Development* 32: 301–309.

Wedgwood, R. 2005. Post-basic Education and Poverty in Tanzania. Working Paper Series No.1. Edinburgh: Centre of African Studies.

6 Gendered Educational Challenges and Achievements in the Tanzanian Context

To build and develop the nation and the well-being of its people, education has been a significant and highly prioritised political instrument in Tanzania. This chapter gives, to begin with, a short historical glimpse into the two major educational reforms in the country, namely Education for Self-Reliance (ESR) and Universal Primary Education (UPE), which commenced following independence in 1961;[1] then, the more recent educational policies are reviewed. In brief, this section leads us from 'Education for Self-Reliance' to 'Education for All' (EFA), as macro-political approaches and initiatives.

Historical Glimpse

The educational policy formulations of post-independence Tanzania have been guided by the principles of *equity, access and quality* manifested early on the adoption of the Arusha Declaration and the ESR in 1967. In accordance with these principles, the notion of *gender* has been part of the policy rhetoric and agenda ever since.[2] ESR aimed to create a literate adult population in four years, whereas the second key point in the history of educational development in Tanzania was the Musoma Resolution to enact UPE in 1974; the UPE efforts were consolidated in 1978 with an Educational Act requiring compulsory school attendance for every child from the age of 7 to 13 (see e.g. Bhalalusesa 2011; Buchert 1991, 1994, 1997; Koda 2007; Sabates, Westbrook and Hernandez-Fernandez 2012; see also Ten/Met (Tanzania Education Network/ Mtandao wa Elimu Tanzania http://www.tenmet.org) and Vavrus 2002b). UPE was implemented and advanced by abolishing school fees, campaigned for by the media, government leaders and the ruling socialist party, and it was organised throughout the entire educational structure from national/ministerial level to regional, district, ward and village levels of the educational establishment. As a result, during the UPE year of 1977/1978, there was a massive increase in student enrolment, with near gender parity, followed, however, by more staggered years (Sabates, Westbrook and Hernandez-Fernandez 2012, 4) Even today, the Tanzanian post-independence commitment to ESR and UPE is reflected in the relatively high adult literacy rate: for instance, in 2006, 69 per cent of adult Tanzanians were literate in comparison to the 63 per cent in sub-Saharan Africa on average; furthermore, 78 per cent of youth (aged 15 to 24) were literate, with a gender parity index of 0.94 (Lugg, Morley and Leach 2007).

Where the abolishment of school fees was adopted as a pivotal strategic tool to reach UPE (by 1977), teacher recruitment, training and deployment were urgently and radically amended to meet the demands of the suddenly evolved mass enrolment. Yet the various adapted measures were not sufficient. Omari and his colleagues have remarked that the government was completely unprepared for the new situation, because the increase in enrolment 'came like thunder' (Omari et al. 1983, 79). To give an idea of the impact of the 'thunder', Malewo, for instance, (1992) has estimated that by the end of the 1970s there was a shortage of 45,000 primary school teachers. It goes without saying that neither the number of schools and the classroom conditions nor the teaching and learning materials were sufficient to accommodate the increased number of enrolled students, and needed to be urgently improved. As pointed out by Sabates, Westbrook and Hernandez-Fernandez (2012), because of the characteristically 'community-centered' ESR, the districts and wards, not to mention the parents of the students, were left with the lion's share of the development of the school infrastructure and the teaching and learning facilities, which paved the way for increasing regional disparities and inequalities in the country; alongside this, the private education sector started to expand (Vavrus 2002a). Furthermore, as might be expected, the quality of education declined rapidly under any measures of learning outcomes. The dilemma between the quantity and quality of education is exactly what the policy-makers are facing and have to contend with today, not only in Tanzania (e.g. Sabates, Westbrook and Hernandez-Fernandez 2012; Towse et al. 2002; Wedgwood 2007), but also in other countries in SSA (see e.g. Hardman et al. 2011; King 2005). As concluded by Sabates and his colleagues (2012; also King 2005 and Wedgwood 2007), the immense growth in student enrolment during the first wave of UPE in the late seventies was possible only at the cost of quality. At the same time, they (ibid.) remind us that what was achieved in such a short period and with such scarce and limited resources was simply astonishing. For example, by the beginning of the 1980s, each village in the country had a primary school and gross primary school enrolment reached nearly 100 per cent.

For all that, the 1980s have been described in general as the 'decade of decline' in Tanzania. As a result of the economic crises in the 1970s, the Government introduced a series of macro-economic reforms in 1981 with support from like-minded bilateral donors (mostly Nordic countries). Despite these 'home-grown' initiatives, the government was persuaded to develop its economic reforms further in agreement with the IMF and the World Bank, the first agreement being signed in 1986 (additional agreements were signed in 1987, 1991 and 1996) in the form of Structural Adjustment Programmes (see Lugg, Morley and Leach 2007). The programmes impacted severely on the macro-economy of the country (see e.g. Tsikata 2001) and the government faced difficulties in financing its social services; from the perspective of educational development, the outcomes deteriorated (see e.g. Brock-Utne 2000). To give two examples, by 1993, gross enrolment in primary education declined from the previously mentioned 100 per cent per cent to 82 per cent, and illiteracy increased between 1986 and 1992 from 10 to 16 per cent. (Alonso i Terme 2002)

In the 1980s and 1990s educational policy reforms in Tanzania intensively followed global patterns in development co-operation in education, substantially defining the strategic visions, objectives and priorities and the operational targets of educational development in the country. These international trends, country case studies and national contexts, including Tanzania, have been reviewed and discussed by Kenneth King and other experts in the field of education and development (King and Buchert 1999); and the Tanzanian experience regarding this particular period is analysed in detail by Lene Buchert (1994, 1997, 1998), just to name a few. One of the key policies that still guide the provision of education in the country is the Education and Training Policy (ETP) formulated in 1995. In addition, the two national macro-policies that define education as a national priority are the Tanzania Development Vision 2025, conceived in 1998, and the more recent National Strategy for Growth and Reduction of Poverty I and II (NSGRP/MKUKUTA), adopted in 2001 and 2010 respectively.

Based on the ETP, the government started to develop the Education Sector Development Programme (ESDP) in 1996, to address the problems with education provision, and to face the new challenges of socio-economic reforms and the increasing demands for human resources development. This is what Sabates, Westbrook and Hernandez-Fernandez (2012) call 'the second wave of the UPE' in Tanzania. A sector-wide approach (SWAP) to education was also initiated to redress the problem of fragmented intervention. ESDPs (1996–2001; 2001–2006) and the current sector programme that is in place until 2017 (URT 2008) aim at operationalising various policies pertaining to sub-sectors of the ETP, which include basic education (consisting, by definition, in the Tanzanian context, of pre-primary, primary, adult, secondary and teacher education), higher education and formal and non-formal vocational education.

In addition to national policy formulations, the Government of Tanzania is a signatory to international and regional human rights agreements asserting equity policies and non-discrimination policies in education. The most relevant and influential that concern education, gender and development, based on universal human rights, are the Millennium Declaration and the World Declaration on Education for All (EFA). Two of the Millennium Development Goals (MDGs), approved by world leaders in 2000 and by the UN World Summit in 2005, refer explicitly to education and gender equality. One of the earliest platforms for the MDG's was precisely the EFA World Conference held in Jomtien in 1990, which was continued by the World Education Forum in 2000, resulting in the Dakar Framework for Action aiming to ensure equal opportunities to education for all.[3] In line with the MDGs and EFA process, the government of Tanzania is committed to equity and non-discrimination policies in education, and has initiated a series of reforms in the education sector to ensure equal access to primary education of good quality for all children, to expand secondary education, and to promote education and empowerment for girls.

Broadly speaking, gender equity and equality is recognised in the current macro-policies in Tanzania as a development goal per se and, moreover, as a cross-cutting issue in sustainable social development and the poverty reduction process, reflecting the commitments discussed above and particularly the ones initiated at the Fourth World Conference on Women, held in Beijing in 1995 (followed by the special session of the General Assembly Beijing+5, and the Platforms of Action Beijing+10 and +15). Furthermore, Tanzania has been bound into international gendered obligations and legal standards since 1980, when the Government signed (and ratified in 1985) the Convention on the Elimination of All Forms of Discrimination Against Women (CEDAW). In addition to the international agreements mentioned above that explicitly cover education and gender, it should be noted that the government of Tanzania is a signatory to the Universal Declaration of Human Rights (UDHR). Although the Declaration is a non-binding resolution, various human rights instruments (e.g. the International Covenant on Civil and Political Rights and the International Covenant on Economic, Social and Cultural Rights) and international treaties (for instance the CEDAW) embedding the rights of the UDHR have acquired the force of human rights law, which binds the signatory states. Furthermore, the profound principles of the UDHR are enshrined in the Constitution of the United Republic of Tanzania (1977; amended in 1984 and 2000). Consequently, the government is explicitly committed to the 'equality and rights of every person' and to the 'equal distribution of resources and social services', implicitly safeguarding the rights of women *and* men, and to *educational* resources and services, among other areas of the social sector.

My study is not about, and does not take part in, the debate concerning the ownership of development and the conditionality of international co-operation (see e.g. Alasuutari and Räsänen 2007; Dengbol-Martinussen and Engberg-Petersen 2003; ECOSOC 2008; Tsikata 2001[4]); neither does it comment on or take a stand on the rationale and justification of Tanzanian policy formulation, strategic priorities and target setting in relation to education, gender and development. Rather, the policy commitments, which the Government of Tanzania is obliged to meet, are comprehended as *facts*, as a macro-structure that pre-conditions and contextualises micro-level functionings and human agency. Accordingly, the 'enabling environment' that contextualises the improvement of girls' and women's positions (as illustrated in Figure 2.1) has been outlined here to show the legal and policy macro-levels of the enabling environment, which aim to 'protect and promote the rights and welfare of women' including gender equality in education. Overtly, this section has been about the historic context and alignment of educational ideologies and policies and the macro-political structures framing the second wave of UPE. Hence, in this brief discussion of the historical education policy backdrop which has led us to the twenty-first century, what is being done in practice to realise the national and international promises with regard to 'Equity Access to Quality Education' and what is the current status of gender equality in education, accomplished within the education sector reforms implemented from the year 2001 onwards?

Politics of Equity, Access and Quality in Education

Today, the overall goal of the education sector development in Tanzania is to ensure the realisation of the principles of equity, access and quality at all levels of education, thus echoing one of the targets of the MDG Goal 3 (to promote gender equality and empower women), which states that by 'no later than 2015, gender disparities are eliminated in all levels of education'. However, as we learned earlier, ever since independence, the focus of education development and reforms in Tanzania has been on primary education. Furthermore, MDGs and EFA Goals have significantly determined the national target setting focusing on primary and secondary education (see e.g. Lewin 2005); thus, in Tanzania, 'eliminating gender disparities' is predominantly targeted at primary and secondary education. Consequently, educational privilege in the country today, as in general in sub-Saharan Africa, is on 'ensuring that by 2015 all children, particularly girls, [...] have access to and complete, free and compulsory primary education of good quality' (see EFA goals in the Appendices; cf. SDGs 4 and 5).

As explained in the introduction (and following the policy–research–practice knowledge construction), the starting point of this research was development-policy driven, guided by the interests and needs of development co-operation in education taking place exactly at the time of the second wave of the UPE in Tanzania in the early 2000s. Thus, to initially gain a (pre-) understanding of the status of education, gender and development in Tanzania, situational analysis was carried out on the basis of policy documentation: to begin with, the national and international education sector monitoring and evaluation reports were collected and analysed. Following this, information gaps and gender challenges within the first implemented education sector development programmes were identified. The policy data under examination in this chapter is mainly from the years between 2001 and 2007 (2010 for some data). Apart from the programme documents (ESDP I 2001–2006, PEDP I 2002–2006, SEDP I 2004–2009 and NSGRP/ MKUKUTA 2001) and the annual performance reports, three sources of macro-level data and policy information, namely the 2005 'Education sector situation analysis' by Roy Carr-Hill and Joyce Ndalichako, the 2005 Poverty and Human Development Report and the 2003–2007 compilation of the Basic Education Statistics in Tanzania (BEST 2009), were crucial in laying the foundation for further micro-level investigations. In addition, The State of the World's Children by UNICEF (2006) on 'the double dividend of gender equality' was an important source of information with which to refine the situational understanding of girls' and women's education and schooling in Tanzania, within the context of sub-Saharan Africa.

In this section I examine the strategic priorities and instruments and the (gendered) achievements and challenges of the programmes listed above. To move beyond basic education, and because the interest of this research is in examining how to reach the highest level of education, a situational analysis of higher education from the gender perspective is briefly carried out. I wish to underline that the main idea behind the analysis is not to focus on gender differences as such, even though I have examined the data as segregated according to gender. I do point out the most evident gender differences with regard to access, participation and attainment, but alongside this, the overall trends concerning female students' attainment and challenges in the Tanzanian education system are scrutinised. The reason for presenting the educational policy strategies and practices in such detail in here is that later on, in the empirical chapters 4, 5 and 6, the same themes are studied thoroughly in terms of the research participants' educational experiences and insights. Later, I will also pay attention to and discuss the very latest (2010–2012) findings from government statistics concerning education sector development in Tanzania from the gender perspective. Here, however, to begin to unravel the phenomenon of girls' and women's education and schooling in Tanzania on the basis of educational policy documentation, the focus is on the data and the timeline that were defined earlier.

Recent Reforms: Primary and Secondary Education

Primary Education of Good Quality

The Primary Education Development Plan (PEDP), introduced in 2001, was the first outcome of efforts to turn international education obligations into feasible strategies and actions within the educational establishment in Tanzania. The Plan pursued both qualitative and quantitative improvements and had four strategic priorities: Enrolment Expansion, Quality Improvement, Capacity Building and Institutional Arrangements. Here, the focus is on the issues of enrolment and quality. Thus, when looking at the two selected strategic priorities of the Plan in more detail through the policy-monitoring lens, what kind of picture can be drawn of educational development and advancement?

In the late 1990s, the net enrolment rate for primary education in Tanzania was 57 per cent per cent (URT 2005). Consequently, the highest priority for the PEDP I was given to *enrolment expansion*, aiming at all children aged 7–13 being enrolled in primary education by 2005. Strategically, this was executed by abolishing school fees and all other mandatory parental contributions and by mobilising communities, particularly through information, education and communication campaigns. Regarding its highest priority, all the reports indicate that PEDP I was a success. According to BEST, in 2009, the gross enrolment ratio (GER) reached 110 per cent and the attained net enrolment ratio (NER)[5] was 96 per cent. Enrolment ratios were equal between girls and boys. Furthermore, gender parity was reported to be practically 1:1 in all the regions (URT 2009a). As a matter of fact, according to the basic education statistics, the gender parity index for primary school enrolment already varied from 0.96 to 1.02 from the mid-90s to 2009.

The generally very positive trend with regard to enrolment was, however, challenged by more analytical reports. In their study covering the entire education sector, Carr-Hill and Ndalichako (2005) pointed out that there were still challenges and concerns regarding regional inequalities (with the NER varying between 88 per cent and 100%), difficulties in enrolling the remaining children and relatively high dropout and repetition rates (see also HakiElimu 2005). After a few years of PEDP I implementation, 76 per cent of the children had reached grade 5 and only 54 per cent had completed their primary education. From a gender perspective, it is worthwhile noting that the GPI for the graduation ratio was 1.01 (Lugg, Morley and Leach 2007). Another noticeable point is that the net attendance rates give a somewhat different picture from the net enrolment rates: for instance, in 2002, the figures were 68 per cent and 81 per cent respectively (URT 2005). There were more than 350,000 repeaters in 2007, without significant differences between girls and boys, yet the total number decreased from previous years. In 2007/08 the repetition rates were highest in Std I-II and Std IV-V and the dropout rate clearly for Std IV–V. In comparison, the promotion rate from one grade to another was the highest in Std VI–VII. The data regarding transition rates in primary school was not arranged according to gender in the 2009 BEST.

In most schools in Tanzania, the significantly increased enrolment rates for Std I resulted in the overcrowding of schools, and children repeating years even aggravated the problem. On top of that, overage enrolment was evident in the Tanzanian primary schools, as verified in the GERs (cf. Lewin 2009). Apparently, repetition and overage enrolment burden all educational resources. According to studies on educational patterns and decision-making, poverty and financial constraints, lack of teaching and learning facilities (e.g. distance to schools, too few schools, poor quality of schools), lack of qualified and competent teachers and parental attitudes are key factors for repetition and dropout (Carr-Hill and Ndalichako 2005; see also Lewin 2007, 2009; Little 2008; and Lugg, Morley and Leach 2007). In primary schools, the main reasons for dropout were truancy (67%), followed by death (5.6%), pregnancy (5.5%) and lack of school requirements (4 per cent). In absolute numbers, there were over 4,000 girls/young women who left school at a very early age (URT 2009a). The government called for an urgent remedy for truancy, though reasons for being absent were neither traced and presented, nor analysed (Okkolin, Lehtomäki and Bhalalusesa 2010).

The second priority of PEDP I was *quality improvement* focusing on three main components: i) improving teacher's teaching styles and methods in the classroom ii) ensuring the availability of good quality learning and teaching materials and iii) ensuring the necessary support for maintaining educational standards aiming at improving the overall achievement of both female and male pupils in the primary education system (URT 2001, 9). Evidently, to improve the quality of primary education, teachers were seen as key agents (see also Towse et al. 2002). PEDP I aimed to enhance teachers' knowledge and competencies by acquiring and developing 'academically sound, child-friendly and gender-sensitive pedagogical skills' (URT 2001, 9–10), but the focus in the implementation of PEDP I was, however, on better conditions at the school level (construction of schools and classrooms, improvement of sanitary and water facilities, purchase of learning materials such as books etc.), and not so much on the subject content or pedagogic capacities of the teachers.

Several micro research studies, but also PEDP I Reviews in 2003 and 2004, which examined classroom teaching, showed that the quality of teaching was poor and that very few changes had occurred in teaching styles or classroom management: teaching was observed to be teacher centred; very few of the teachers used textbooks during the lessons; pupils were passive recipients of transmitted information from teachers; punitive measures were still taken against children; and the lack of basic awareness of the holistic nature of child-centred and active learning approaches was evident among teachers, head teachers, ward education coordinators and district education officers. This is in line with the conclusion drawn by Hardman, Abd-Kadir and Tibuhinda (2012), who reviewed studies relating to classroom pedagogy in sub-Saharan African countries, including Tanzania, which are, as they point out, altogether comparatively few in number. One example from Tanzania regarding teacher training and pedagogical practices is an ethnographic study of a teacher training college carried out by Francis Vavrus (2009). The governmental reports suggested that, in principle, the participatory teaching and learning methods were welcomed by many teachers but in practice, they lacked either the materials or support when it came to actual adoption and implementation (URT 2003; URT 2004a).

With regard to the strategic priority of quality improvement, the Plan proposed recruiting a certain number of new teachers. More teachers were recruited than originally planned, but there was still a shortage, which meant that PEDP I targets for recruitment were underestimated. PEDP I aimed at establishing teacher-to-pupil ratios (TPRs) that could accommodate the increases in enrolment rates, and efforts were made to ensure an equitable and gender-balanced distribution of trained teachers. In 2009, the national average ratio was 1:54, though large regional variations existed, ranging from 1:42 in Dar es Salaam to 1:70 in Shinyanga (URT 2009a). The government re-estimated that it would take five years to clear the backlog of teacher demand in primary education. However, as pointed out in the HakiElimu review (2005), the government reports did not factor in the numbers of teachers dying from HIV/AIDS, for instance, or retiring. To tackle the issue of a lack of qualified teachers further, one of the PEDP I objectives was to improve the use of the teachers already employed. One of the strategies was to increase the teacher–pupil contact time through effective teacher management; another strategy was to ensure effective multi-grade and double-shift teaching; thirdly, it was stipulated that the minimum school week should contain twenty hours of instruction time. According to HakiElimu (2005) there was no evidence, however, as to whether these activities had taken place in practice.

It is important to note the significance that the teacher–pupil ratio had on the examination results: in general, those regions with a below average TPR tended to have below-average examination results, and vice versa. As an example, the Std IV Primary School Examination pass rates were the highest in Dar es Salaam (97%) in comparison to 79 per cent in Shinyanga (compare the TPRs above), although the lowest pass rate was attained in the Morogoro region, at 70 per cent of candidates. During the first year of PEDP I implementation, the deployment of teachers in and between different regions and districts remained problematic. The problem was particularly acute in remote areas where teachers were unwilling to be posted: better-qualified teachers tended to be found in urban areas; and female teachers were concentrated in urban schools. This related partly to the lack of adequate facilities, notably houses, and partly to the provision of suitable incentives. Incentives and housing conditions in remote areas are one of the recommendations made by Ruth Wedgwood (2005) when she studied post-basic education and poverty in Tanzania. These practicalities were raised, among other things, by Towse and his colleagues (2002) as the rationale for non-graduate students to choose or not choose a teaching career in Tanzania. Besides, surveys of the living and working conditions of teachers have shown the importance of housing, respect and an adequate salary paid on time for teachers to remain in their profession – and how these have actually altered for the worse over the years (HakiElimu 2005; Carr-Hill and Ndalichako 2005). According to the comparative study of six sub-Saharan African countries by Education International (2007), teachers' salaries were lowest in Tanzania and Gambia, being notably higher in the other countries. The allure of the teaching profession was also problematised by the participants in this research, as we will hear from their narratives later.

One of the capacity development objectives of the PEDP I that focused on teachers concerned training an adequate number of qualified teachers to meet the demand, and to ensure that new teachers had appropriate pedagogical skills and knowledge. This was done by upgrading the qualifications, knowledge and skills of teacher training college tutors, and re-organising the content and delivery of pre-service teacher training[6] (URT 2001). As pointed out by Towse et al. (2002), the average academic qualification for those entering primary teacher training in Tanzania is low, since they have normally completed four years of lower secondary school and have graduated with an ordinary level (O-level) secondary education certificate. When it came to the recruitment of new teachers, the government decided to reduce student training at the teacher training colleges (TTCs) from two years to one, and during the second year, student teachers were meant to be posted to schools and receive on-the-job training (e.g. Wedgwood 2005). Mentor teachers, for their part, were supposed to receive professional support for their guidance and tutoring. According to HakiElimu (2005, 24), in practice, the student teachers received very little support while in the field, and the mentor teachers were somewhat reluctant and were not highly motivated. To enhance the quality of the rapidly expanding teaching force, the Tanzanian Institute of Education, which centrally determines the curriculum for teacher education in Tanzania, began to develop and revise curricula to be implemented in 2005. Substantial efforts were made in the content and delivery of the pre-service teacher training curriculum, but despite its achievements, namely, the incorporation of participatory methods and cross-cutting issues, it had shortfalls: in brief, as pointed out by Carr-Hill and Ndalichako (2005; see also URT 2004a), repeating the key findings of various previous reports and studies, the main concern was encapsulated in the definition and meaning of the 'quality of education'.

The second of the *quality improvement* components of PEDP I outlined the importance of having an adequate supply of teaching and learning resources, and considerable efforts were made to improve the learning environment (desks, latrines, water facilities etc.) and the availability of learning materials (textbooks, exercise books and other learning materials, e.g. science kits). Although the targeted pupil-to-book ratio of 3:1 was not achieved in 2002, the situation improved from previous years, and parents were reported as being happy with the development (URT 2003). As with the teacher-to-pupil ratio, there were variations in the pupil-to-book ratios between and within districts and also between different subjects. Furthermore, the Public Expenditure Tracking Study (PETS) found that 'the leakage' of funds sent to School Councils was largest for the purchase of books. The desk situation was reported to be critical (five children to each desk) and the situation regarding water was even more critical (HakiElimu 2005; URT 2005). In general, only 62 per cent of the Tanzanian people have access to improved water facilities, with great variation between urban (85 per cent) and rural (49 per cent) populations (UNICEF 2006). In order to improve the learning environment and to meet other expenditures, all schools were allocated a capitation grant and required to open bank accounts to deposit the money. The PETS found that the disbursement of the grant from the central administration to the district level had been in line with the initial plans, but the recorded inflow of overall capitation grant to the school level ranged between 54 and 64 per cent. In other words, about 40 per cent of the resources did not reach the school level (in time or at all) (HakiElimu 2005).

The Plan also initiated programmes to build sufficient classrooms and sanitary facilities to accommodate the expanded student enrolment. There were substantial school construction activities during the PEDP I, and construction of new classrooms proved to be a major success of the Plan. Despite the speed and numbers of classrooms, the quality of workmanship was reported to be questionable (Carr-Hill and Ndalichako 2005). To illustrate the issue of sanitation: according to UNICEF (2006), a total of 47 per cent of Tanzanians (53 per cent in urban and 43 per cent in rural areas) had access to adequate sanitary facilities;

according to BEST (URT 2009b), only 38 per cent of schools had adequate sanitation facilities with an average ratio of 1:60 (unisex latrines), with huge regional disparities in the latrine coverage, ranging from 24 per cent in Shinyanga to 60 per cent in Kilimanjaro and Iringa. Finally, apart from school construction, PEDP I stated that teachers' houses were to be constructed to provide incentives, especially to female teachers, to work in remote rural areas (URT 2001). Reports showed, however, that the number of teachers' houses constructed had fallen well below the target, meaning that only 20 per cent of the planned construction was actually completed (URT 2004a).

The emphasis placed on teachers, both in terms of numbers and their qualifications, as well as the importance given to sufficient school infrastructure and learning environments, suggests that these aspects are presumed to have a great impact on student performance and overall educational achievement. During the PEDP I implementation there were changes in student performance. In 2006, 80 per cent of the candidates passed the Std IV Examinations, which was 9 per cent less than three years previously; in the following year, the percentage dropped again, but two years later, it rose to 85 per cent. The trend in the Primary School Leaving Examination (PSLE) in Std VII was the opposite: in 2000, 22 per cent of the students passed, in 2002, the rate was 27 per cent, in 2004, 49 per cent and in 2006, it rose to 71 per cent of the candidates; however, in 2008, the pass rate practically collapsed and only 53 per cent of students passed the exam (URT 2009a). The 'peak' of the PEDP I reform (and the ability of the government to cope with the increase in demand) and its impact on educational achievement, which just might take time to settle, interestingly runs parallel to the phenomenon discussed by Sabates, Westbrook and Hernandez-Fernandez (2012) in relation to the first UPE wave in Tanzania (see also Wedgwood 2005).

From the gender perspective, there was a relatively big divide between female and male performance with regard to the Primary School Leaving Examination (PSLE): 46 per cent of female students in comparison to 60 per cent of their male counterparts passed the examination in 2008. The highest pass rate for girls was reached in 2006 (65 per cent), but the trend has been declining ever since (URT 2009a). The 2004 Joint Review of PEDP (URT 2004a) found that a gulf existed between the rhetoric on what needs to be done, and the reality as regard to what is actually being done to address gender issues. However, in the light of more recent reports and analysis regarding access, participation and attainment in primary education as a whole, Tanzania is doing rather well, and, fairly, the Tanzanian initiatives and experiences have been applauded (see Okkolin, Lehtomäki and Bhalalusesa 2010). In contrast to the almost complete enrolment in primary education, access to secondary education is extremely limited in Tanzania. In addition, from the gender perspective, participation and achievement is even more limited: in brief, the boys both outnumber and outperform the girls (see e.g. Lugg, Morley and Leach 2007, URT 2009a, URT 2009b). The next stage of analysis relates to secondary education.

Secondary Education

The importance of secondary education for sustainable socio-economic development and for personal benefit has been recognised in education sector policy formulation in Tanzania. The potential of secondary education for social and individual return, in contrast to the econometric studies emphasising the importance of primary education, was also evidenced in a study by Wedgwood (2005) examining post-basic education and poverty in Tanzania. As in other sub-Saharan African countries, the Tanzanian secondary education system has faced serious challenges due to the changes in society, the economy and the labour market. Moreover, the massive increase in enrolment rates in primary education called for the expansion of secondary education. Therefore, the Secondary Education Development Plan (SEDP 2004–2009) followed PEDP I and was embraced in 2003/04.

The overall goal of the SEDP I (URT 2004b) was to increase the proportion of Tanzanian youth completing secondary education with 'acceptable outcomes'. The secondary education sector in Tanzania has been one of the smallest in sub-Saharan Africa, constituting only 10 per cent of the total enrolment in basic education (primary education being 82 per cent), whereas the equivalent figure in the rest of SSA has been 20–30 per cent (e.g. Kenya 23 per cent and Namibia 34 per cent in 2004) (URT 2010b). The NER in secondary education has increased every year since 2004, from as low as 8 per cent to 13 per cent in 2006, and again to 28 per cent in 2009. As noted by the government, there was an increase in total enrolment by 180 per cent from 2005 to 2009 (URT 2009a). It is important to note, however, that when looking at Forms I–IV (O-level) and Forms V–VI (A-level) separately, the ratios in 2009 were 30 and 1.5, respectively (ibid.) Hence, regardless of the significant increase in the transition rate from primary to secondary education, the number of students who actually continue further to the advanced level of secondary education is extremely low. One of the concrete previous bottlenecks was that there were simply not enough schools to accommodate students eligible for secondary education, but during the implementation of SEDP I, the number of schools more than doubled (URT 2010b). Yet the number of new schools and classrooms has not boosted the number of students and urged them to continue further.

Alongside the overall expansion of secondary education, and as a result of the affirmative action policy to have equal numbers of female and male students in secondary education, the number of females increased during the implementation of SEDP I (URT 2009b). In 2010, with regard to access, the quantitative differences between females and males seemed relatively small, and the proportion of female students in Form I was 45 per cent. Although enrolment at the entry level of secondary education was close between female and male students, in Forms V and VI, the female proportions were 40 per cent and 39 per cent, respectively (URT 2010a). In comparison to the first years of SEDP I implementation, when the gender parity index in the final year of secondary education was 0.5 (i.e. the relation between male and female students was 2:1), there was development for the better; however, the declining trend in retention of female students is apparent and does not meet international and national policy objectives.

Furthermore, female students' underperformance at lower secondary level has increased the gender gap at the higher levels of education. One of the SEDP I objectives was to improve the quality of secondary education, and, more particularly, to reduce the poor performance in the national Form IV and VI examinations. The students' level of performance improved during SEDP I, but the gender imbalance in performance and achievement is one of the major aspects that the Ministry of Education and Vocational Training (MoEVT) still needs to address (URT 2009b). For instance, the percentage of female students passing the Form II, IV and VI examinations was significantly lower than males throughout the whole of the 2000s: in 2008, only 65 per cent of female students passed. It needs to be noted that of those females who reached the final year of secondary education and sat the exam, 99 per cent did pass, which is 3 per cent higher than the rate of their male colleagues. However, what has to be remembered is that only a tiny proportion of female students enter the final Form VI level. An interesting result was obtained from the pass rates divided according to single-sex and co-educational schools: in 2005 both sexes attained over a 99 per cent pass rate in single-sex schools, whereas the pass rate for males in co-educational schools was 63 per cent, and for females this was as low as 33 per cent (URT 2009b). This is also an issue that the research participants will point out and discuss in their narratives concerning their secondary school experiences.

During the first years of SEDP I implementation, the dropout rates increased and were the highest in Form II (where an examination was introduced). In 2009, truancy was the leading cause of reported dropouts in secondary education, as in primary education, followed by pregnancy and lack of school requirements. This means that the schooling and the construction of an educational career were discontinued (permanently?) for nearly 5,000 girls because a policy of readmission had not yet been established (at the time of my research). Another aspect to be considered is that, as indicated by UNICEF (2006), 49 per cent of rural Tanzanian women and 23 per cent of their urban counterparts are married before they turn 18. The official age to enter secondary education is 14. The difference in fertility between women with and without secondary education is large, but the gap between rural and urban Tanzania is even larger (Wedgwood 2005). Altogether, a significant number of Tanzanian women get married and establish a family right at the time of their official secondary school age, bearing in mind that only 1.5 per cent of Tanzanians reach the advanced level of secondary education, of which the proportion of female students is less than a half. On that basis, it is evident that only a minimal number of women in Tanzania progress to the next level of education and reach the higher education levels.

Gender and Higher Education

> Tanzania should be a nation with '[a] high level of education ...; a nation which produces the quantity and quality of educated people sufficiently equipped with the requisite knowledge to solve the society's problems, meet the challenges of development and attain competitiveness at regional and global levels'.
>
> (URT 1999a)

The Development Vision 2025 (URT 1999a; see also URT 1999b) cited above holds that it is pivotal for the Tanzanian people and the nation to move beyond basic education, since having people equipped with only primary and secondary education is not sufficient to face contemporary social and societal demands and challenges. At the time of envisaging the future for the country in 1999, there were around 6,500 students in the whole country enrolled for university degree programmes, representing barely 0.2 per cent of Tanzanians (URT 2010b). Ten years earlier, Tanzania had only 3,146 students attending its two universities, which was less than one-tenth of the number in Kenya, for instance (Mkude, Cooksey and Levey 2003). More recently, gross enrolments have risen, reflecting, among other factors, the overall increase in the numbers of students completing basic education: in 2008/2009 there were nearly 96,000 students engaged in university-level studies and the GER was reported to be 3.0 (URT 2010a).

Despite the expansion of tertiary education worldwide, including Tanzania, only a relatively small number of the relevant age group has access to this level of education. To give two examples: in 2006, the GER for higher education in sub-Saharan Africa was 5 per cent, the lowest regional GER in the world, in contrast to the highest at 70 per cent in North America and Western Europe; as a result of the high participation in the global north, a child entering primary school can expect to spend three years of their life in higher education, whereas the corresponding amount for a child in sub-Saharan Africa (out of those 5 per cent reaching higher education) is less than half a year (Lugg, Morley and Leach). In the sub-Saharan context, the Tanzanian experiences with regard to the growth in enrolment have been both slow and low alike. To give some figures, in 2007, the GER in higher education in Tanzania was 2 per cent, and in neighbouring Kenya and Uganda, for instance, the rate was 3 per cent; in West Africa, in Nigeria the GER was 10 per cent, and South Africa had reached a GER of 15 per cent (URT 2010b). In short, regardless of increasing the speed and the growth, the GER in tertiary education was extremely low in Tanzania. Furthermore, as argued by Ishengoma (2004), the provision of higher education remained elitist (see also Kotecha, Wilson-Strydom and Fongwa 2012; Wilson-Strydom 2015).

As pointed out by Lugg, Morley and Leach (2007, 11), the participation rates for women in higher education increased between 1999 and 2004 in all regions of the world. Yet in several regions, including sub-Saharan Africa, the GPI remained notably low. Globally the gender parity index was 1.03, suggesting that participation rates were higher for women than for men, but in the context of sub-Saharan Africa, the GPI was 0.62 in 2004 (ibid.). In Tanzania, gender imbalance is listed explicitly as one of the major problems in the current Higher Education Policy (URT 2010b). In 2010, the GER for higher education was 5.3 and the proportion of women 35 per cent (including Bachelors, Master's and Doctoral degree students). Hence, only a third of university students in Tanzania were women, and the increase in the GER in Tanzanian higher education, following the trend in SSA, has been male dominated (URT 2010b).

As pointed out earlier, apart from gender *differences* as such, pre-eminent concerns to be elaborated on here are the overall *patterns* regarding female students' access, retention and attainment within the education system in the country. Consequently, the axiomatic challenge is that within the education system there is also a declining trend in the **proportion** of female students, as is the case in educational attainment. The widening gender gap in female students' educational progression is clearly exhibited in the figures from 2006, presented in the ministerial Strategic Plan for Gender Mainstreaming (URT 2009b):

- **49 per cent** of students enrolled in Primary Std I were female and 45 per cent passed the final exams;
- **47.5 per cent** of first-year secondary school students were female and 44 per cent completed the O-level and gained the certificate of secondary education examination (only 26 per cent of female candidates who sat the exam got the first Division, the highest grade);
- **40.5 per cent** of the enrolled students at A-level were female and 31 per cent of female students passed the exam (however, only 35 per cent **entered the examinations**);
- **30 per cent** of Master's degree students were female – among others, the first four women participating in this research.

To sum up, and remembering that the GER for higher education in 2006 was extremely low (only 2 per cent), it is clear that only 'one in a million' Tanzanian women reached the highest level of education.

Another concern in higher education, having its roots in basic education, is the small proportion of female students in science-based programmes: in 2007/2008 the participation rate for female students in these subjects was about 10 per cent (URT 2010b). This seems to be a somewhat universal pheno-menon and concern, and in Tanzania, it is explicitly raised in national policy formulation and targeting. The patterns of participation in higher education programmes, based largely on disciplinary and subject choices during the students' earlier educational career, were also part of our conversations with the research participants and will be presented in the women's narratives later. Briefly, on the one hand, the subject choices for the research participants were first and foremost guided by their overall school performance, but on the other hand, by the familial and broader social expectations and assumptions embedded in the *idea of Tanzanian woman*. The former is explicated in the current Higher Education Development Programme (URT 2010b), which states that performance in science subjects at A-level examinations, from which the pool of female students qualifying for science-based programmes are picked, is low, as is the size of the whole pool (see also Mbilinyi 2000). The latter is polemised by Mbilinyi (ibid.) in describing the various factors and perspectives impacting on the overall admission for higher education from the gender perspective, and especially in terms of maths and science-related disciplines. Interestingly, Mbilinyi (ibid.) included the Musoma Resolution as one of the explanations having a negative impact on the proportion of female students in higher education. However, in contrast to the common apprehension and jargon, she concluded that the low female ratios in science-related subjects were already over-emphasised; instead, attention should be given to increasing enrolment in *all* subjects. The University of Dar es Salaam, as a case studied by Mkude, Cooksey and Levey (2003) and Mbilinyi (2000; see also Mbilinyi and Mbughuni 1991), along with governmental commitments and objectives, has initiated various schemes and practical measures to promote female students joining science-based programmes, but more importantly, to increase women's overall admission to higher education. These initiatives have included earmarked financial aid, pre-entry programmes, lowering entry cut-off points and bonus points, giving women priority in campus accommodation, gender-sensitisation programmes etc. (e.g. Lihamba, Mwaipopo and Shule 2006; URT 2010b) Hence, the political will to promote and achieve gender equality at 'all levels of education' in Tanzania is operationalised in various ways and at different levels of the educational establishment.

Yet it is important to acknowledge that affirmative actions do not tackle gender inequality alone, but that they may permeate through a range of intersecting dimensions of educational disadvantage such as social class, religion and ethnicity (e.g. Mkude, Cooksey and Levey 2003; Wedgwood 2005). Nevertheless, quite often it is exactly the poor female students coming from remote and less developed regions and districts, no matter how high a potential with regard to education they might have, who are excluded from education. For this reason, the more profound and far-reaching social and human aspects of education are increasingly emphasised and discussed by researchers in the field of education, gender and development (see e.g. Arnot and Fennell 2008b; Colclough 2008; Fennell and Arnot 2009; Leach 2000; Morley and Lugg 2009; Robeyns 2006a; Unterhalter e.g. 2003a, 2007a).

'Why Haven't More of These Actions Been Implemented?'

To sumarise, the main notion made on the basis of the information presented in the previous section is that the second wave of the UPE significantly emulated the first, massive post-independent education reform of the 1970s. For that reason, what was done, what has been achieved, and what remains a challenge in the twenty-first century is remarkably consistent with what has already been experienced, and, as rather ominously noted by Williams (2005), there is a strong sense of déjà vu. Citing and agreeing with many other Tanzanian writers, Ruth Wedgwood (2005, 28) has claimed that the push for UPE between the late 1970s and early 1980s was the major cause of the deterioration of quality at all levels of education, even impacting on the education system today. Next, I follow their explanation concerning the perpetuating cycle of poor education progression in the country.

As we learned, the abolition of school fees was the major factor in striving for UPE in Tanzania, supported, for instance, by information campaigns. This encouraged Tanzanian parents to send their children to school, and as a result, massive enrolment took place. However, the education system was totally unprepared in the 1970s, as it was in the 2000s, to accommodate the increased numbers of students, and to respond to their (and their parents') needs and expectations. Consequently, all educational resources proved to be both insufficient and inadequate: a lack of (qualified) teachers was the major problem, and with regard to the school environment, there were not enough schools in the first place, and classrooms were extremely overcrowded; there were not enough basic facilities such as desks and chairs; there was a lack of teaching and learning facilities, e.g. books, and the water and sanitation facilities proved to be extremely poor and limited, to name the most evident deficiencies. Evidently, such school and classroom conditions advanced neither pedagogically sound teaching nor good and meaningful learning. Strategically, as a reaction to high demand, teacher recruitment, training and deployment were prioritised and reorganised. This meant, first, lowering the minimum qualification requirements, and second, shortening teacher training. This, together with the poor learning environment at schools, worsened the already poor learning results. Alongside this, absenteeism, repetition and dropout rates remained high and even increased, as parents and students themselves lost faith in schooling and the value of education. In addition, other concrete implications of the poor groundwork at the primary level of education were the very low secondary education transition and enrolment rates, and those students who actually reached secondary school faced somewhat similar school environments to those of the primary schools. Consequently, the retention rates and performance remained poor, and yet this was the pool from which to draw students for higher education. Apparently, the higher education sector in Tanzania faced serious problems and challenges.

What then was achieved? Tanzania came very close to achieving UPE in the early 1980s, as is the situation now, and according to some statistics, Tanzania will reach UPE by 2015 along with the MDG and EFA Goals. Also in secondary and higher education, significant increases in total enrolment took place. However, according to Wedgwood's (2007, 383) examination, Tanzania is one of the most pertinent examples of a country where the efforts to get all children enrolled in primary education yielded little apparent benefit in the long run. Wedgwood's (2005, 2007) claims concern particularly the connection between education and poverty reduction; but she suggests that the quantitative targets in Tanzania were achieved at the cost of quality, which is mirrored in the low achievements in secondary and higher education, particularly if measured with more quality targeted indicators (e.g. school attendance versus school enrolment, and survival, completion and transition rates). This was also concluded in the review by Carr-Hill and Ndalichako (2005) and reports by HakiElimu (2005, 2010).

From the gender perspective, the status of the education sector in Tanzania reflects the overall patterns of development. The declining trend in both quantity and quality is prominent, meaning that the higher you go, the fewer female students you have: this is partially due to low educational performance (because of various reasons that will be discussed in the next section), but partially due to the expectations and assumptions attached to the *idea of Tanzanian woman*, including the subject choices and level of education that are seen as appropriate for female students. There were some explicit strategies to promote gender equality and equity in the PEDP I and the SEDP I, such as increasing the number of qualified female teachers (only 29 per cent of enrolled students in Teacher Training Colleges in 2006 were female) (URT 2009a), but instead of precise strategic means with which to tackle gender imbalance, more emphasis was given to *gender mainstreaming* 'at all levels of education'. In short, gender mainstreaming refers to institutionalising gender justice by transforming (constraining) social structures, which can relate to both socio-cultural meanings and institutions alike (see e.g. Goetz 1997). Swainson (2000) has identified how the key development agencies operating in Tanzania in the education sector have employed different approaches to promote gender equality. Her case study claims that for instance, UNICEF redresses gender equality with gender-specific initiatives, whereas DFID, for instance, has consistently integrated women/gender into its overall approach, that is, mainstreamed gender. However, changing social meanings and institutions requires concrete tools and instruments to be implemented. As remarked by Swainson (ibid.), policy recommendations need to be backed up by concrete plans of action, and furthermore, structural arrangements are required to ensure that the gender aspect is integrated into both policy and practice. Factors that impact on gendered progression, as well as on the overall advancement of the education sector, are intertwined in many ways, and for that reason, they are difficult to hold onto. This was noted by Marjorie Mbilinyi in a paper in 2000, in which she refers to and repeats a report she conducted together with Patricia Mbughuni and others ten years earlier, providing a detailed list of suggestions of actions to be implemented to achieve gender equity and transformation. Mbilinyi admitted that it was not possible to implement 100 actions or more (at the same time), but that prioritising is necessary; still, she wondered 'Why haven't more of these actions been implemented?' (Mbilinyi 2000, 27) An identical question was posed by Nicola Swainson (2000) in a paper comparing the design and implementation of gender policies in Tanzania, Malawi and Zimbabwe. In particular, she pointed out how all three countries had significant gender components in their educational policies, yet how little progress had been made. To give one example, she mentioned how the recommendation for the re-admittance of pregnant school-girls was already given in the 1995 Education and Training Policy, but how it was brushed aside by the all-male team of senior MoEC officials who approved the final draft of the ETP (Swainson 2000). As discussed above, re-admittance has now been incorporated again in the draft of the revised the ETP; however, this has not yet been approved.

I have pointed out that factors that impact on female students' educational advancement are intertwined in various and complex ways. For example, there is an obvious linkage between school attendance and performance, but in education sector monitoring, attendance rates were not reported as intensively and seen as important indicators of progress as were the enrolment rates, for instance. Yet absenteeism was reported to be the main reason for repeating and dropping out, but the reasons for not attending school were not collected, presented and analysed. Furthermore, attending school is not only about schooling, but includes the pivotal aspects of social and human development cited above, and intersects critically with the legal, economic, political, social and familial levels of the enabling environment. In other words, the reasons and reasoning for pursuing education are constructed throughout the educational establishment; and from the gender perspective, in accordance with the *idea of Tanzanian woman*, which includes also the idea of quantity and quality of education. Although the policy data does not examine what is behind the figures and uncover the reasons behind truancy, for example, on the basis of previous research on gender, education and development we have learned about the meanings and institutions that prohibit and/or enable female students' education and schooling. Next, the gendered meanings and understandings of equality and equity in education are looked at and analysed through the capabilities lenses.

Notes

1 The United Republic of Tanzania was formed in 1964 as a result of the union of the mainland of Tanganyika and the archipelago of Zanzibar.

2 By definition, (gender) *equity* (e.g. in access to resources, goods, services and decision-making) denotes the equivalence in life outcomes for women and men, recognising their different needs and interests, and requiring a redistribution of power and resources; it means fairness and impartiality in the treatment of women and men in terms of rights, benefits, obligations and opportunities. Hence, the essence of equity is not identical treatment: treatment may be equal or different, but should always be considered equivalent in terms of rights, benefits, obligations and opportunities. (Gender) *equality*, distinctively, denotes women and men enjoying equal rights, opportunities and entitlements in civil and political life, including the ability to participate in the public sphere (including education). The terms of 'gender equity' and 'gender equality' have been also conceptualised as substantive equality and formal equality, respectively.

3 The origin of the Goals can be traced to UN conferences and other international summits of the 1990s, which drew up a number of key development goals and targets. The earliest platforms for the MDGs were the World Conference on Education for All, held in 1990, and in the same year, the International World Summit for

Children. The other foundational conferences and summits for the MDGs were those of Rio de Janeiro in 1992 on the environment, Rome in 1992 on nutrition, Vienna in 1993 on human rights, Cairo in 1994 on population and development, Copenhagen in 1995 on social development, Beijing in 1995 on women, Istanbul in 1996 on human settlements, Rome in 1996 on food and Hanoi in 1998 on the 20/21 initiative. The MDG exercise may then be seen either as a consolidation of various (previously unachieved) goals (see e.g. Saith 2006), or as a visionary 'framework for development and time-bound targets by which progress can be measured' (UN 2006, 3).

4 'Ownership' is one of the core principles of the Paris Declaration (OECD) http://www.oecd.org/dac/effectiveness/parisdeclarationandaccraagendaforaction.htm

5 NER = enrolled children in the official school age group/total number of children in the official school age group; GER = enrolled children of all ages/total number of children in the official school age group; GPI = ratio of females to males (e.g. in enrolment).

6 The teacher education system in Tanzania: Primary education: Grade C Certificate holders are entitled to teach in the first two grades of primary education; a Grade B Certificate is obtained by promotion or by completing four years of training at a teacher training college after grade seven; Grade A Certificate holders are entitled to teach in all levels of primary education after completing a two-year course at the end of Form IV at lower secondary (O-level). Secondary education: Work experience and a two-year Advanced Diploma programme after Form VI is required to teach at O-level; upper secondary (A-level) teachers are required to hold a post-graduate diploma or Bachelor's degree.

Bibliography

Alasuutari, H., and R. Räsänen. 2007. Partnership and Ownership in the Context of Education Sector in Zambia. In *Education Sector Programmes in Developing Countries. Socio-political and Cultural Perspective*, ed. T. Takala, 115–163. Tampere, Finland: Tampere University Press.

Alonso i Terme, R. 2002. The Elimination of the Enrollment Fee for Primary Education in Tanzania: A Case Study on the Political Economy of Pro-Poor Policies. Paper presented at the Joint Donor Staff Training Activity/Partnership for Poverty Reduction, June 17, in Tanzania.

Arnot, M., and S. Fennell. 2008b. (Re)visiting Education and Development Agendas: Contemporary Gender Research. In *Gender Education and Equality in a Global Context: Conceptual Frameworks and Policy Perspectives*, ed. S. Fennell and M. Arnot, 1–16. London and New York: Routledge.

Bhalalusesa, E.P. 2011. 'Education For All' Initiatives and the Barriers to Educating Girls and Young Women in Tanzania. *Papers in Education and Development* 30: 104–132.

Brock-Utne, B. 2000. *Whose Education for All? The Recolonization of the African Mind.* New York: Falmer.

Buchert, L. 1991. *Politics, Development and Education in Tanzania 1919–1986: An Historical Interpretation of Social Change.* PhD diss., University of London.

Buchert, L. 1994. *Education in the Development of Tanzania 1919–1990.* Athens, OH: Ohio University Press.

Buchert, L. 1997. *Education Policy Formulation in Tanzania: Coordination Between the Government and International Aid Agencies.* Paris: UNESCO.

Carr-Hill, R., and J. Ndalichako. 2005. *Education Sector Situation Analysis: Final Draft Report Revised.* Dar es Salaam, Tanzania: Economic and Social Research Foundation.

Colclough, C. 2008. Global Gender Goals and the Construction of Equality: Conceptual Dilemmas and Policy Practices. In *Gender Education and Equality in a Global Context: Conceptual Frameworks and Policy Perspectives,* ed. S. Fennell and M. Arnot, 51–66. London and New York: Routledge.

Dengbol-Martinussen, J., and P. Engeberg-Petersen. 2003. *Aid: Understanding International Development Co-operation.* London: Zed Books.

ECOSOC. 2008. *Report of the First Development Co-operation Forum.* New York: ECOSOC.

Education International. 2007. *Teacher Supply, Recruitment and Retention in Six Anglophone Sub-Saharan African Countries.* Brussels, Belgium: EI/EFAIDS.

Fennell, S., and M. Arnot. 2009. Decentring Hegemonic Gender Theory: The Implications for Educational Research. RECOUP Working Paper No. 21. London, UK: DFID.

Fukuda-Parr, S., C. Lopes, and K. Malik. 2002. *Capacity for Development: New Solutions to Old Problems.* New York: UNDP.

Goetz, A.M. 1997. *Getting Institutions Right for Women in Development.* London: Zed Books.

HakiElimu. 2005. *Three Years of PEDP Implementation: Key Findings from Government Reviews.* Dar es Salaam, Tanzania: HakiElimu.

HakiElimu. 2010. *Education in Reverse: Is PEDP II Undoing the Progress of PEDP I? Brief No.10.1E.* Dar es Salaam, Tanzania: HakiElimu.

Hardman, F., J. Ackers, M. O'Sullivan, and N. Abrishamian. 2011. Developing a Systematic Approach to Teacher Education in Sub-Saharan Africa: Emerging Lessons from Kenya, Tanzania and Uganda. *Compare: A Journal of Comparative and International Education* 41, no. 5: 669–684.

Hardman, F., J. Abd-Kadir, and A. Tibuhinda. 2012. Reforming Teacher Education in Tanzania. *International Journal of Educational Development* 32: 826–834.

Ishengoma, J.M. 2004. Cost sharing in Higher Education in Tanzania: Fact of Fiction? *Journal of Higher Education in Africa* 2, no. 2: 101–133.

King, K. 2005. Re-targeting Schools, Skills and Jobs in Kenya: Quantity, Quality and Outcomes. *International Journal of Educational Development* 25: 423–435.

King, K., and L. Buchert, eds. 1999. *Changing International Aid to Education: Global Patterns and National Contexts.* Paris: UNESCO.

Koda, B. 2007. Education For Sustainable Development: The Experience of Tanzania. Paper presented at UniPID/EADI Symposium on Accessing Development Knowledge-Partnership Perspective, April 16, in Helsinki, Finland.

Kotecha, P., M. Wilson-Strydom, and S.N. Fongwa. 2012. *A Regional Perspective.* Vol. 1 of *A Profile of Higher Education in Southern Africa.* Johannesburg, South Africa: SARUA.

Leach, F. 2000. Gender Implications of Development Agency Policies on Education and Training. *International Journal of Educational Development* 20: 333–347.

Lewin, K.M. 2005. Planning Post-primary Education: Taking Targets to Task. *International Journal of Educational Development* 25: 408–422.

Lewin, K.M. 2007. *Improving Access, Equity and Transitions in Education: Creating a Research Agenda.* No. 1 of *CREATE Pathways to Access Series.* Brighton, UK: CREATE.

Lewin, K.M. 2009. Access to Education in Sub-Saharan Africa. Patterns, Problems and Possibilities. *Comparative Education* 45, no. 2: 151–174.

Lihamba, A., R. Mwaipopo, and L. Shule. 2006. The challenges of affirmative action in Tanzanian higher education institutions: A case study of the University of Dar es Salaam, Tanzania. *Women's Studies International Forum* 29: 581–591.

Little, A.W. 2008. *Size Matters for EFA.* No. 26 of *CREATE Pathways to Access Series.* Brighton, UK: CREATE

Lugg, R., L. Morley, and F. Leach. 2007. *Widening Participation in Higher Education in Ghana and Tanzania: Developing an Equity Scorecard. Country Profiles for Ghana and Tanzania: Economic, Social and Political Contexts for Widening Participation in Higher Education.* London, UK: DFID.

Malewo, A. 1992. *Ualimu Katika Shule za Msingi (Teaching in primary schools).* Dar es Salaam, Tanzania: University Press.

Mbilinyi, M. 2000. *Gender Issues in Higher Education and their Implications for Gender Mainstreaming and Strategic Planning.* Dar es Salaam, Tanzania: UDSM Gender Dimensions Programme Committee.

Mbilinyi, M., and P. Mbughuni, eds. 1991. *Education in Tanzania with a Gender Perspective: Summary Report.* Dar es Salaam and Stockholm: SIDA.

Mkude, D., B. Cooksey, and L. Levey. 2003. *Higher Education in Tanzania: A Case Study.* Oxford, UK: James Currey.

Morley, L. and R. Lugg. 2009. Mapping Meritocracy: Intersecting Gender, Poverty and Higher Educational Opportunity Structures. *Higher Education Policy* 22, 37–60.

Morley, L., F. Leach, and R. Lugg. 2009. Democratising Higher Education in Ghana and Tanzania: Opportunity Structures and Social Inequalities. *International Journal of Educational Development* 29: 56–64.

Okkolin, M.-A., E. Lehtomäki, and E. Bhalalusesa. 2010. Successes and Challenges in the Education Sector Development in Tanzania: Focus on Gender and Inclusive Education. *Gender and Education* 22, no. 1: 63–71.

Omari, I., A.S. Mbise, S.T. Mahange, G.A. Malekela, and M.P. Besha. 1983. *Universal Primary Education in Tanzania.* Ottawa, Canada: International Development Research Centre. http://idl-bnc.idrc.ca/dspace/bitstream/10625/7084/1/IDL-7084.pdf.

Robeyns, I. 2006a. Three Models of Education: Rights, Capabilities and Human Capital. *Theory and Research in Education* 6, no. 1: 68–84.

Sabates, R., J. Westbrook, and J. Hernandez-Fernandez. 2012. The 1977 Universal Primary Education in Tanzania: a Historical Base for Quantitative Enquiry. *International Journal of Research & Method in Education* 25, no. 1: 1–16.

Saith, A. 2006. From Universal Values to Millennium Development Goals: Lost in Translation. *Development and Change* 37, no. 6: 1167–1199.

Swainson, N. 2000. Knowledge and Power: The Design and Implementation of Gender Policies in Education in Malawi, Tanzania and Zimbabwe. *International Journal of Educational Development* 20: 49–64.

Towse, P., D. Kent, F. Osaki, and N. Kirua. 2002. Non-graduate Teacher Recruitment and Retention: Some Factors Affecting Teacher Effectiveness in Tanzania. *Teaching and Teacher Education* 18, no. 1: 637–652.

Tsikata, Y.M. 2001. Owing Economic Reforms. A Comparative Study of Ghana and Tanzania. WIDER Discussion Paper No. 2001/53. Helsinki, Finland: UNU-WIDER.

UN. 2006. *The Millennium Development Goals Report.* New York, United Nations.

UNICEF. 2006. *The State of the World's Children 2007. Women and Children. The Double Dividend of Gender Equality.* New York: UNICEF.

Unterhalter, E. 2003a. Crossing Disciplinary Boundaries: the Potential of Sen's Capability Approach for Sociologists of Education. *British Journal of Sociology of Education* 24, no. 5: 665–669.

Unterhalter, E. 2007a. *Gender, Schooling and Global Social Justice.* London: Routledge.

URT. 1999a. *Development Vision 2025.* Dar es Salaam, Tanzania: Planning Commision.

URT. 1999b. *Higher Education Policy.* Dar es Salaam, Tanzania: Ministry of Science, Technology and Higher Education.

URT. 2001. *Primary Education Development Plan 2002–2006.* Dar es Salaam, Tanzania: Ministry of Education and Culture.

URT. 2003. *Joint Review of the Primary Education Development Plan (PEDP): Final Report.* Dar es Salaam, Tanzania: Ministry of Education and Culture.

URT. 2004a. *Joint Review of the Primary Education Development Plan (PEDP): Final Report.* Dar es Salaam, Tanzania: Ministry of Education and Culture.

URT. 2004b. *Secondary Education Development Plan 2004–2009.* Dar es Salaam, Tanzania: Ministry of Education and Culture.

URT. 2005. *Poverty and Human Development Report 2005.* Dar es Salaam, Tanzania: Government of Tanzania.

URT. 2008. *Education Sector Development Programme (ESDP) 2008–2017.* Dar es Salaam, Tanzania: Ministry of Education and Culture.

URT. 2009a. *Basic Education Statistics in Tanzania (BEST) 2003–2007 National Data.* Dar es Salaam, Tanzania: Ministry of Education and Culture.

URT. 2009b. *Medium Term Strategic Plan for Gender Mainstreaming 2010/11–2014/ 15.* Dar es Salaam, Tanzania: Ministry of Education and Vocational Training.

URT. 2010a. *Basic Education Statistics in Tanzania 2006–2010.* Dar es Salaam, Tanzania: Ministry of Education and Vocational Training.

URT. 2010b. *Higher Education Development Programme 2010–2015: Enhanced Relevance, Access and Quality in Higher Education.* Dar es Salaam, Tanzania: Ministry of Education and Vocational Training.

Vavrus, F. 2002a. Making Distinctions: Privatisation and the (Un)educated Girl on Mount Kilimanjaro, Tanzania. *International Journal of Educational Development* 22: 527–547.

Vavrus, F. 2002b. Uncoupling the Articulation Between Girls' Education and Tradition in Tanzania. *Gender and Education* 14, no. 4: 367–389.

Vavrus, F. 2009. The Cultural Politics of Constructivist Pedagogy: Teacher Education Reform in the United Republic of Tanzania. *International Journal of Educational Development* 14, no. 1: 65–73.

Wedgwood, R. 2005. Post-basic Education and Poverty in Tanzania. Working Paper Series No.1. Edinburgh: Centre of African Studies.

Wedgwood, R. 2007. Education and Poverty Reduction in Tanzania. *International Journal of Educational Development* 27: 383–396.

Williams, P., 2005. Universal primary education for a second time in Anglophone Africa. In Beveridge, M., King, K., Palmer, R., Wedgwood, R. (eds.) *Reintegrating Education, Skills and Work in Africa.* Centre of African Studies, University of Edinburgh, Edinburgh, pp. 65–84

Wilson-Strydom, M. 2015. *University Access and Success: Capabilities, Diversity and Social Justice.* New York: Routledge.

7   Capabilities Lenses to Address and Assess Education, Gender and Development

As discussed previously, the 'equality of capabilities' approach has affinities with the 'distributive equality' and particularly the 'equality of conditions' approaches to *understand*, and consequently to *pursue* and *assess* gender equality in education. Notwithstanding that, the capability approach (CA) is foundationally distinctive, as will be explained through a brief outline of the relevant CA literature.

Core Idea and Key Concepts

As may be inferred from the very terminology, it is *human capabilities* that are to be equalised within the CA in pursuing gender equality in education, and not (only) the resources and/or social conditions and arrangements. This is not to claim, however, that the resources are of no significance; quite the contrary. Resources (such as the number of classrooms, books and the teacher—pupil ratios) are important as a means to well-being and achieving equality (Sen 1999, 73), but not as an end; 'the end' and the ultimate goal in the capability approach is each person, and, to be more precise, the well-being of every human being (Sen 1985a; Crocker and Robeyns 2010). Nor does the CA claim that social conditions and contexts, interrelationships between people and singular circumstances, are irrelevant; quite the opposite. Individual well-being is advantaged or disadvantaged via macro-level policy environments (such as civil and political rights, cf. economic, social and cultural rights, and non-discriminatory educational policies[1]), but, just as importantly, through micro-level social structures (which can be institutions, everyday life practices and meanings alike).

Correspondingly, the assessment of well-being within the capability approach is based on the various beings and doings that people can *achieve*, on the one hand, and the *opportunities* or substantive *freedoms* to realise those beings and doings, on the other. According to the vocabulary of the CA, these are functionings and capabilities respectively (Sen e.g. 1992, 39–53; 1993, 30–53; 1999, 75). That is to say that *functionings*, to signify the various valuable beings and doings, and *capabilities*, understood as an opportunity concept of freedom (Sen 1985b, 3–4; 2002; see also Kaufman 2006), are the constituent elements of the CA, and the relationship *between* the achievement/outcome and the opportunity/potential to achieve is one of the key aspects of analysis within the approach.

In order to illustrate the core idea of the CA, arising from the distinction between functionings and capabilities, I provide an example of two young girls, who *both fail* in mathematics in an international study of learning achievements, following from on Walker and Unterhalter's (2007, 5) initial case from Kenya. Suppose two Tanzanian girls enrolled in the final grade of their primary education, one attending a well-equipped school in Arusha with qualified and motivated teachers offering a conducive and safe learning environment, while the other comes from Shinyanga, one of the educationally (statistically) worse regions in the country. For the former, the major reason for failing was her decision to spend less time on mathematics and more time with friends in drama club and on other leisure activities; for the latter, despite her interest in mathematics and schoolwork in general, the failure was mainly due to the lack of a mathematics teacher; she was actually taught by an English specialist. For her, private tuition to study after school was available in principle (as for the first girl), but in practice, her parents could not afford this for all of their children and decided to give preference to their son. Instead she was allocated household chores, leaving very little time to study at home and prepare for the examinations. Thus, it is evident that if the assessment of educational advancement and well-being is based on the functionings (only), we see an equal outcome (failure); if, however, we assess the girls' capabilities to function, we end up with an inequality induced by the girls' different sets of capabilities. Therefore, according to the CA, we should look beyond the actual functionings to the *opportunities* people have in order to function, or, as suggested by Robeyns (2011), we should analyse which *sets of capabilities* are open, for example, to the latter Tanzanian girl described above. Could she simultaneously study well and support the well-being of her family, or is she constrained to choosing between these two functionings, both of which she may value, because they are not simultaneously open to her?

Sen has defined four conceptual spaces which are equally important and interdependent aspects of human life (see Table 7.1). It is important to acknowledge that, whatever concept of advantage one chooses to analyse, the information base of the judgement must relate to it: *well-being achievement* should be measured in functionings and *well-being freedom* is reflected by a person's capability set; as for focusing on *agency*, it transcends the analysis beyond functionings and capabilities alone (Robeyns 2005). One's *agency achievement* refers to the realisation of the goals and values a person has a reason to pursue, whether or not they are connected with his or her well-being, whereas *agency freedom* refers to what the person, as a responsible agent, is free to do and achieve in pursuit of whatever goals or values he or she regards as important (Sen 1985a, 203–204; Sen 1992, 56). Sen emphasises that people's well-being, whether functionings, capabilities or both, need not necessarily embody their motivations and/or objectives (Sen 1985a, 1992; see

Table 7.1 Conceptual Spaces to Evaluate Human Life – Wema's Story.

	WELL-BEING	AGENCY
ACHIEVEMENT	*Well-being Achievements* (= functionings) State of a person: various things one manages to do and be (may be the outcome of one's own or other people's decisions and actions) e.g. complete primary education, continue to secondary education and university, Bachelor's degree in law	*Agency Achievements* Realisation of goals and values one personally has a reason to pursue (may be or may not be connected with one's own well-being) e.g. to study further and get a Master's degree in sociology
FREEDOM	*Well-being Freedoms* (= capabilities) Genuine opportunities and alternatives to function e.g. access to school, library, private lessons, mother's (and grandparents) moral and financial support	*Agency Freedoms* Freedom to set goals and act accordingly; to make choices and decide (as responsible agent) e.g. whether the opportunity to choose not to study law is available

also Alkire 2005, 2008; Samman and Santos 2009; World Bank 2013). A person is exercising agency when she/he acts on purpose and for a purpose, goal or reason. Such an activity is sometimes defined as reasoned agency (Crocker and Robeyns 2010, 61–63; cf. McNay 2000, 2003).

The conceptual spaces and spheres for evaluating educational advancement are illustrated in Table 7.1 and are exemplified by one of the research participant's (Wema, 32) experiences. Wema's parents are divorced, and she and her younger sister and older brother (who passed away in an accident) were raised by their mother, without any support from the father. Two of their cousins also lived with them. The (extended) family was assisted by an uncle and grandparents. Wema's mother held a Master's degree and was working as a teacher. Wema explained how her mother had invested everything she could afford to ensure the children's educational advancement. For her primary education, Wema attended a government school, performed exceptionally well and was selected to join the government secondary school, where education was free and other costs were relatively low. For higher education, she was selected to study law. Studying at university and gaining a place in one of the most competitive fields and in one of the most respected faculties in the country was not difficult for an excellent student such as Wema. Still, she felt it was not what she really wanted. She later decided to move abroad and study for a Master's degree in social sciences, which was compatible with her own aspirations. At the time of the research conversation, she was 32, married and finishing her degree. On the surface, and if assessed on the basis of functionings (alone), her formal educational career manifests as successful, easy and linear.

Vaughan (2007, 117–119) has followed Sen's (1985a; 1993, 30–39; 1999, 36–38) four conceptual spaces within which human life can be evaluated, and made a distinction as per the cross-cutting dimensions of achievement and freedom within the educational sphere. In her educational analysis, *well-being achievement* (functioning) means 'full participation in formal education', including attending school, participating in lessons and learning (basic literacy and numeracy), being somewhat synonymous with Subrahmanian's (2005) 'formal equality' (i.e. access to and participation in education). In comparison, *well-being freedom* (capabilities) in Vaughan's analysis in the educational context denotes one's ability to choose to attend, to participate and to learn, and thereby includes the opportunities and constraints to this freedom, such as tuition and household chores. In turn, 'to choose' implies at first that there are at least two alternatives to be chosen from, and second, that there is someone who decides. This leads to the crucial counterpart of well-being in Sen's thinking; namely, to the aspect of *agency* in human life (Sen 1999, 11, 281–288), which, like well-being, has the dimensions of achievement and freedom. As suggested by Subrahmanian (2005), reaching substantive freedom requires reaching beyond equality of treatment and opportunity and perceiving the cruciality of the aspects of agency and autonomy in human life. With the concept of agency, Sen (1999; cf. 1992; see also Alkire 2008[2]) emphasises that people's well-being, whether functionings, capabilities or both, need not necessarily satisfy one's motivations and/or objectives, but instead, one may pursue goals that may reduce one's well-being (e.g. drama club and leisure instead of educational well-being). In contrast, one's '*agency achievement* refers to the realisation of goals and values she has *a reason* to pursue, whether or not they are connected with one's own well-being' (Sen 1992, 56), whereas '*agency freedom* refers to what the person is free to do and achieve in pursuit of whatever goals or values he or she regards as important' (Sen 1985a, 203–204). In other words, agency achievements refer to, for instance, the particular level of education (e.g. Master's degree), the particular level of attainment (not scoring an F in mathematics) and particular subject choices (e.g. social sciences instead of law) etc., which are, by definition, educational aspects that a person has reason to value; in turn, agency freedom means one's ability to so decide and to act accordingly (Vaughan 2007, 117–119; see also Crocker and Robeyns 2010, 61–63).

Making an interpersonal comparison between the two Tanzanian girls in accordance with the four conceptual spaces discussed above illustrates the different portrayal of equality in education arising from the different information bases. Consequently, both of the girls achieved the same level of educational well-being, which in this case was marked as an 'F' in the exam (functioning = well-being achievement); however, their opportunities to learn mathematics were different because of the school environment, quality of teaching and their household obligations (capabilities = well-being freedom).

With regard to agency, presumably the first girl did not value mathematics but held a preference for drama club and leisure, while the latter valued and was interested in mathematics and schoolwork in general; hence, in relation to their well-being achievement (scoring 'F'), their agency achievements were unequal, and with regard to their potential to decide, that is to exercise agency freedom, for one reason or another, the first girl was able to act according to her objectives and motivations, that is, to hold a preference for issues other than learning mathematics (maybe she was simply not interested in mathematics or maybe she wanted to become an actress, and drama club would assist her in pursuing that goal; if this were the case, then it would again provide a very different portrayal of her well-being, agency, achievement and freedom; maybe her parents did not value compulsory education highly or maybe they agreed with her regarding her acting goals and did not force her to study more; maybe she was in a position to claim that if she were obliged to spend more time on her schoolwork instead of drama, she would drop out school for good). However, the second girl, for one reason or another, had a constrained negotiation position (the preference given to the educational well-being of her brother, although we do not know if this increased his agency achievement or freedom) and hence, a minimum freedom of agency to be exercised for a better well-being achievement.

Indeed, having the negotiation power, and/or other means and resources to be converted into functionings, takes us to another bundle of concepts within the capability approach. As noted earlier, the CA values *inter alia* the physical environments (such as schools, roads) and resources (such as books, chairs and qualified teachers) as the means for achieving well-being. To be more exact, the CA is interested in looking at the relationship between the resources people have and what they can do with them (Unterhalter and Brighouse 2007, 75). Hence, to achieve the various beings and doings that people have reason to value, they are influenced, enabled or constrained by the *conversion factors*, grouped into three according to their sources by Robeyns (2005) and exemplified by the use of a bicycle. As she points out, we are not (necessarily) interested in the bike in itself, as an object, but we (have reason to) value a bike because of what we can do with it; that is, gain mobility (cf. Sen 2002, 43):

- Personal conversion factors (such as sex, age, physical condition):

If a person is disabled or has never learned to cycle, then the bike will be of limited help in enabling the function of mobility.

- Environmental conversion factors (such as geographical location, logistics):

If there are no paved roads (to the school, for instance), then the bike will be useless for the function of mobility.

- Social conversion factors (social and societal policies, practices, norms, values, hierarchies, regarding gender, age, class etc.) – 'the idea':

If the social environment imposes, for example, the idea that women are not allowed to cycle in the first place or when not accompanied by a male family member, then the bike will not enhance the opportunities to function.

Quoting Sen (1999, 76), Tao (2013, 3) notes that if the conversion factors which block the capability freedom can be reconciled, a person would then be judged to have an expanded capability, and her well-being would be evaluated either based on the opportunities reflected in her capability set, or on the functionings that she chose to realise from this set. Recalling, for example, the second Tanzanian girl, who failed her mathematics exam: if her parents were to allow her to go for extra evening classes to prepare for her mathematics exam (after finishing her share of the household chores with the rest of them being allocated to her brother), it would expand her opportunities (capability set) to function; she might take this opportunity, or she might decide not to take advantage of it, being aware of the risk of being harassed in school or on her way back home, and thus this aspect would diminish her capability of succeeding in education (cf. Unterhalter 2003b). Apparently, she is again facing the dilemma of choosing between the two functionings, both of which she may value (school achievement and personal safety), because they are not open to her simultaneously; however, she would be exercising her agency freedom in deciding by herself and acting accordingly.

Thus, to assess *gender equality in education*, the CA argues that, if we evaluate only the functionings (e.g. the achieved grade in mathematics) and do not look at the conditions of choice for the learners, we gain an inadequate understanding of people's well-being, and the policy initiatives are unlikely to be sufficient. Hence, educational institutions (from macro-policies to school-level practices) need to enhance people's functionings *and* capabilities, *and* support people's agency. Correspondingly, strategic policy priorities and practices should be assessed not only on the basis of their impact on people's functionings, but, just as importantly, on the basis of their influence on people's capabilities (e.g. the set of opportunities to achieve a valued grade in mathematics) and their freedoms to act 'in line with his or her conception of the good' (Sen 1985a, 206). To summarise, the capability approach claims that whatever concept of advantage one chooses to consider, the information base of the judgement must relate to that: well-being achievement should be measured in functionings and well-being freedom is reflected in a person's capability set; as for focusing on agency, it transcends the analysis in terms of functionings and capabilities (Robeyns 2005, 102–103). Indeed, the capability approach was initiated and conceptualised by Sen and Nussbaum (e.g. 2000; Nussbaum and Sen 1993) and further developed by a growing number of scholars (see e.g. Robeyns 2006b, 2011) as a response to the 'incapability' of various approaches in measuring and evaluating well-being, and in making interpersonal comparisons (not only between individuals, but between social groups and nations) on those grounds.

The CA is commonly conceived as an approach, rather than an explanatory theory. Moreover, the approach provides a normative framework, rather than a precise theory of well-being, yet for some it is more like a theory of justice (as for Nussbaum) or like a theory of human development (see e.g. Alkire 2010; Robeyns 2006b). For some, in the educational context, it appears to be a categorical tool (Vaughan 2007), yet to be complemented with 'an explanatory theory' such as the critical realist theory of causation (Tao 2013) or Dewey's educational theory of experience (Lessmann 2009). For some, capabilities concepts resonate with the hermeneutics of Arendt and Ricoeur (Nebel and Herrera Rendon 2006), and others may extend the CA through the analytic tradition of phenomenology (Ballet, Dubois and Mahieu 2007). Robeyns (2011) has identified three kinds of normative exercises dominating the capability approach. First, the assessment of individual well-being, second, the evaluation and assessment of social arrangements, and third, the design of policies and proposals about social change in society, all of them making two fundamental normative claims: i) that freedom to achieve well-being is of primary moral importance, and ii) that freedom to achieve well-being is to be understood in terms of people's capabilities (see also Robeyns 2005, 94–96; Robeyns 2006b, 353; Sen 1992, 48). Robeyns (2011, also Crocker and Robeyns 2010) has also remarked on how scholars and policy-makers have used the approach in a wide range of fields, and in a narrow and a broad way, representing a metric for interpersonal comparisons and an alternative framework for evaluation of well-being respectively. Moreover, the capability approach embraces a high level of conceptual abstraction, and this is partially the reason for some philosophical disagreements about the best description of it (Robeyns 2011). Evidently, the approach offers a flexible, open-ended and multi-purpose framework.

In my study, the approach is not conceived of as a 'capability theory' to explain girls' and women's education and schooling in Tanzania; instead, it is conceived of as an alternative framework, whose core elements and concepts are employed to develop an understanding of the research participants' educational experiences and insights; that is, their well-being and agency. Consequently, the analyses of this examination may be located within the first and second of the 'normative exercises', if Robeyns' divisions are followed and somewhat broadly understood. The applicability of the CA's conceptual nexus in the field of education, which has relevance for this study in discussing some critical issues that need to be discussed, such as compulsory education versus freedoms, children, 'real values' versus 'adapted preferences', *voices* of the women etc., is explicated in the next chapter. Prior to this methodological discussion, a brief summary of the understandings and approaches to assessing gender equality in education is in order.

Equality of What?

So asked Sen in the Tanner Lecture on Human Values in 1979. To answer, 'gender parity' is the mainstream method of monitoring and evaluating gender, but as suggested by the very concept, it only measures the equality of amounts and numbers, hence is the 'first-order outcome indicator' (Subrahmanian 2005). Partly because of the lack of a definition of gender equality, the measurement of progress is difficult and subject to disagreement. One common debate concerns, for example, whether equality and social justice (in education) should be assessed in terms of resources or welfare (e.g. Clayton and Williams, in Unterhalter 2007b, 98). To move beyond this polarisation, the capability approach extends to shift the argumentation between resources and achievements towards the 'interpersonal comparisons in the space of capabilities' (Unterhalter 2007b, 99; cf. Sen 1992, 50). Another reason for the difficulty in assessing gender equality in education is that, different from the 'distributive equality' approach, the 'equality of condition' and 'capabilities approaches' unfold the scope and mandate of education policies and schooling, operating in intersecting policy arenas, and, more importantly, in the private spheres of human beings concerning immensely subjective matters. Evidently, the 'stories of gender equality' in education narrated on the grounds of the two latter approaches are very different from the gender parity stories (which arise from the presuppositions and methodology embedded in the distributive equality approach).

In my study, all three understandings and approaches are utilised for different purposes. To begin with, building on the methodology espoused and applied by the government of Tanzania, gender equality as parity has been examined, at first, to gain a 'pre-understanding' of the phenomenon of girls' and women's education and schooling in Tanzania, and second, to 'build the case' to be further investigated (see Moser 2007; Reinharz 1992). Thus, the processes to pursue, and the achievements and challenges concerning 'the formal equality' (Subrahmanian 2005) that is 'access to' and 'participation in' education, which the government of Tanzania is committed to in its non-discriminatory policy formulations, were reviewed to commence this research. However, gender equality only rests on, but is not the same as, achieving gender parity (through equal distribution), as remarked on by Subrahmanian (2005) and Unterhalter (2007a; 2007b). Hence, for the purpose of this research, this understanding of gender equality is insufficient. It is plausible to claim that the comprehension of gender equality as parity is also inadequate for the government of Tanzania, to achieve the equality goals in education to which it is nationally and internationally tied. Presumably, in accordance with the rights-based rhetoric employed in education policies in the country, gender equality signifies something more akin to substantive equality rather than equal amounts, hence recognising the structural, constructed and relational nature of women's disadvantagement, and bringing onto the agenda the human development and qualitative aspects of education and schooling, teaching and learning. However, referring to the Mbilinyi report (1991) and paper (2000) and the review conducted by Carr-Hill and Ndalichako (2005), there seems to be a 'gap' between rhetoric and reality in education sector development in Tanzania.

The challenge embedded in the understanding of substantive gender equality in education is that the methodology employed to tackle gender inequalities in education presupposes a knowledge and understanding of the complex relationship between structure and agency, referring to enabling and/ or constraining structures, which constitute both the pre-condition and the outcome of people's agency. The positions taken in the structure–agency debate at the very core of the sociology of education can be grouped into two broad streams. One has focused on how schools reproduce inequalities and social injustice through maldistribution and silencing, by paying attention to educational outcomes and 'functionings'[3]; and because of the influence of sociologists (including me) working in this research area, who mainly vouch for the second kind of understanding of gender, much of the interest has been directed to examining gendered power relations and socio-economically and socio-culturally defined processes and practices in education, including school and classroom practices. In turn, the other group of researchers (including me) has looked more at the conditions in schools and/or other 'enabling environments' through which the learners can contest and/or reshape unequal social relations and practices, being more interested in the process dimensions of justice[4] and the negotiated meanings of equality for individuals (Unterhalter 2007b, 89; Walker and Unterhalter 2007, 7). If this streaming is followed, my study may be defined as the latter approach, and this orientation, which places a particular emphasis on human well-being and agency, resonates with and is amplified in the capability approach. However, instead of methodological individualism, at the core of the approach is ethical individualism (Robeyns 2005, 108), this being one of the main reasons why I, in deciding upon a set of concepts and core elements functional for the purpose of this examination (cf. Moser 2007), ended up positing the study within the broad framework and discussion of the CA. For all that focus on functionings and capabilities and an emphasis on the normative well-being and agency of human beings, which in this study refers to the availability of alternatives and opportunities (structurally given freedoms), decision-making processes, and the question of 'who decides' (reasoning-based freedoms), the other reasons behind the use of the CA relate to the attention it gives to resources and social structures (institutions, practices and meanings), which are conversion factors and contexts *par excellence*, and finally, to the processes through which human beings reach (and realise) their valued beings and doings. Next, the practical adaptation of the approach in my analysis is explained in more detail.

Analytical Adaptation of the Approach in the Study

To recap, the capability approach was initiated in the first place as a response to the inadequacy of various approaches in evaluating and understanding people's well-being, for which reason, the CA asserts, many policy initiatives are unlikely to be sufficient and meaningfully targeted. Unterhalter has illustrated the implication of the weak 'information base' concerning educational policy formulation as per Sen's criticism of income-based welfare economics by arguing that it is 'impossible to develop policy on schooling, for example, as one cannot know how schooling is valued by all individuals in a particular society; how schooling is ranked relative to other important goods, for example that girls remain at home and help with housework rather than go to school; or whether aspirations for or against schooling are the "real" views of an individual, or the views they had to adopt because of powerful customs that dictated appropriate behaviour' (Unterhalter 2007b, 98; cf. Kelly 2012). These notions are at the core of my study, the key assumption being that the women have had a reason to value education and, hence, 'to be educated'; indeed, the women valued and ranked education highly, because they had upgraded their professional qualifications (diploma teachers) not only once (to BA), but twice (to a Master's degree).[5] The second presupposition suggests that the socio-economic and socio-cultural environments of the women have had a great impact on the realisation of 'to be educated'. Thus, in examining these social surroundings and arrangements, the aim is to identify how the women's educational aspirations were valued and ranked by their schools and social and familial environments, and how the environments either constrained or enabled and enhanced their 'opportunities to be educated'. At the core of the study is also an examination of the processes of 'becoming educated',[6] that is, positing one's own aspirations in relation to *the idea of Tanzanian woman*. As remarked by Walker and Unterhalter (2007), the opportunity to imagine ways of being and to act accordingly, are pivotal for positive social change.[7] Hence, the focus of the assessment of educational well-being is both in enabling environments and between resources and achievements. Thus, agency and freedom to make up one's mind about education and schooling as a valued end, and to convert one's aspirations into valued achievements, lies at the heart of this examination (cf. substantive equality) – this being the key characteristic distinguishing the CA from the other approaches, as previously discussed (Unterhalter 2007b, 99).

The conceptual plurality of the CA (as with the vagueness of the definition of gender equality, for instance) gives rise to multidimensionality in terms of measurement in a positive sense, but also to inaccuracy, by definition. Unterhalter, for example, has criticised how Sen in his writings comprehends education as a social opportunity (*capability*), as a valuable outcome (*functioning*), and as a causality of *freedom* (as part of the process of exercising *agency*) implying that there is confusion over the different meanings of education (Unterhalter 2003). Furthermore, education is, for Sen, one of the very few *basic capabilities*, and generally speaking, the practitioners of the CA regard education as a basic capability, 'irrespective of their stance on spelling out a complete definition of well-being in the space of capabilities' (Biggeri 2007, 197; see also Hart 2009). On the other hand, as outlined by Terzi (2007), the capability *to be educated* rather than education in itself can be considered as a basic capability, first, because the lack of this opportunity would essentially harm and disadvantage the individual, and second, the capability to be educated plays a substantial role in the expansion of other intersecting and constitutive capabilities. Hence, education can be understood as a complex good that entails both intrinsic and instrumental values. This idea is mirrored in Vaughan's (2007) study on measuring capabilities, in which she divides the capabilities to participate in and capabilities gained through formal education respectively. Similarly, Walker (2006, 2007) points out the intrinsic worth of education and its instrumental value and potential, in applying the capability approach to analysing gender equality in South Africa (by interpreting what a group of South African girls say matters in their lives). In line with this, Saito (2003) discusses this dual aspect of the CA in relation to education, this being close to 'equality within' and 'equality through' education, as discussed within the gender equality discourses (Subrahmanian 2005).

The intrinsic and instrumental value of education is explicitly articulated in the Tanzanian Education and Training Policy (URT 1995; under revision at the time of my study) maintaining that 'education is a process by which the individual acquires knowledge and skills necessary for appreciating and adapting to the environment and the ever-changing social, political and economic conditions of society as a means by which one can realize one's full potential'. The ETP (URT 1995) claims further that 'a good system of education in any country must be effective on two fronts: on the quantitative level, to ensure access to education and equity in distribution and allocation of resources to various segments of the society, and on the qualitative level, to ensure that the country produces the skills needed for rapid social and economic development'. The above quotes are suggestive of: first, the respect of education for human development *par excellence*; second, its benefit for the society at large; and third, the responsibility of the government to establish, and to assure the reachability of, such a space of capabilities (cf. Nussbaum 1997, 2000, 2004). In my study, the intrinsic value of education, understood somewhat as a basic capability, is acknowledged, in accordance with the rights-based rhetoric and positioning endorsed in the Tanzanian and international policy formulation; however, I have been more interested in the research participants' capabilities 'to be educated', and in that regard, the position taken here approximates to Terzi's position. However, I have not paid attention to the research participants' reasoning in terms of 'to be educated', as to whether they have attached intrinsic, instrumental, or indeed both attributes to education, due to the ontological comprehension of human agency, as explicated previously.

'Reasoning' is an issue that many of the capability scholars pay attention to when discussing the applicability of the approach in assessing the well-being and agency of children, who may not value education, let alone going to school. Saito (2003), for example, refers to Sen, who has stated that while it may be the case that children's freedoms are constricted under compulsory education, the importance of schooling in nurturing future capabilities and freedoms ought to be considered and given priority; in comparison to formal, compulsory education, the picture would be different for adult or higher education, presuming that being non-compulsory and with the curriculum generally chosen by the student, it involves a potentially much higher level of well-being and agency freedoms (Vaughan 2007). Biggeri (2007), together with his colleagues (2006, 2012), has identified at least five issues to be taken into account when working with children's capabilities and agency.[8] One concerns the aspect of intrinsic value and the instrumental nature of education discussed above; and another the fact that children's capabilities are at least partially affected by the capability set and achieved functionings of their parents. The third observation is linked to the possibility of converting one's capabilities into functionings, being dependent on the decisions of parents, guardians and teachers (whose desires and interests and freedoms are served?), and this being intertwined with the degree of autonomy in the process of choice (at different ages of the child). The fourth aspect concerns the age and the life cycle of the child, in which regard Nussbaum (2003 in Walker and Unterhalter 2007) is clear in asserting that children should be required to remain in compulsory education until they have developed the capabilities that are important in enabling them to have genuine and valued choices. The last remark made by Biggeri (2007, 199) concerns the role of children in building up future society; that is, their role in contributing to shaping the future conversion factors, for which reason Biggeri et al. (2006) have claimed a higher priority for children's capabilities. Hence, in addition to *intrinsic* importance and *instrumental* contributions, the significance of agency parallels its *constructive* role in the creation of values and norms (Alkire 2008), resonating with the social constructionist comprehension of change by reconstructing social structures and dispositions. What Biggeri brings to the fore is of importance for my study, since much space is given to the women's childhoods and early school experiences: first, because primary education is the level of education in Tanzania where much policy emphasis and prioritising occurs and resources are invested; and second, because early school experiences seem to impact on later educational aspirations and advancement (e.g. URT 2010a; Okkolin, Lehtomäki and Bhalalusesa 2010; see also Burchardt 2009). On the other hand, reaching higher education refers often to the choices and decision-making of an adolescent, and in most of the women's cases, of an adult, and thus to a 'potentially much higher level of well-being freedoms and agency freedoms' (Vaughan 2007, 128).

For all that, in studying children's *voices* and educational aspirations, Unterhalter (2012) has aimed at paying attention to the possibility that, indeed, children adapt and express what they think is possible. What she (ibid.) claims is not to diminish the importance of children's voices concerning preferences, even adapted preferences,[9] in policy-making and assessments; yet she stresses the consequential role of the context in formulating preferences (adapted or not), and thus, the importance of locating 'voices' and 'pathways' (cf. Subrahmanian 2005). However, it is critical to note the remark made by Clark (2009) that the adaptation argument may be used to undermine the moral case for listening to the voices of the poor in his example, applicable to children, disabled people and girls and women alike (e.g. Malhotra and Schuler, 2005; Narayan 2002; Nussbaum 2000, 2001), who may be regarded, generally speaking, as marginalised and disadvantaged groups, as silenced and often excluded from the decision-making processes concerning their own lives (e.g. Clark 2003, 2009; Kabeer 1999, 2005a, 2005b; cf. Lehtomäki 2005). In my study, to gain a more complete and meaningful comprehension of girls' and women's education and schooling in Tanzania, including the critical aspect of educational well-being and agency (capability approach), the policy-research perspectives are complemented with the practice perspective (hermeneutics); that is, the *voices of the women* concerning their capability sets, which, in turn, relates to their school and social and familial environments; that is, their educational contexts. This is the point where the Berger and Luckmann kind of constructionist paradigm (discussed in Chapter 4), which directs the interest to the language and 'voices' to examine people's beings and doings and becomings (compare Gadamer 1975, xxii), and towards social everyday life and customary practices, meets some of the key elements that the capabilities scholars aim to reveal.

In analysing the linkages between adaptation, capability sets and agency goals, Burchardt (2009, 16) has suggested that 'any approach that places value on individual freedom must define and protect a sphere within which individual choices are respected.' In the same way, in outlining the conception of basic capabilities, Alkire has given weight to the fundamental element of choice, its relation to the pursuit of people's valuable ends and objectives, and hence to their well-being (Alkire 2002). Thus, alongside the conceptual jungle of education in terms of i) functioning, ii) capability, iii) part of the capability set to be converted into functioning, iv) basic capability etc., another area of confusion arises concerning the conception of freedom and agency, intertwined with the concepts of adaptation and autonomy; that is, the debate on to what extent human beings are capable of making rational judgements and choices, and to act accordingly in a particular social setting (e.g. Alkire 2002; Clark 2009). Stewart (1995, in Robeyns 2006b), for instance, in contrast to most capabilities theorists and practitioners, has proposed including the exercise of choice as one of the relevant functionings. In turn, for Sen, human agency refers to the ability to act on behalf of goals that matter to individuals, and people have to be seen 'as being actively involved – given the opportunity – in shaping their own destiny, and not just as passive recipients of the fruits of cunning development programmes' (Sen 1999, 53). In short, according to Sen's approach, the expansion of opportunities (capabilities), together with the expansion of process freedoms (exercise of agency), is what actually defines development. This is also what Subrahmanian and Unterhalter mean by the move beyond opportunity, and including the aspects of agency and autonomy, to reach substantial gender equality in education. To continue with the multiplicity of the concepts and measures of agency in Sen's writings, Alkire has not taken on the various characteristically different features such as a) process freedoms and b) opportunity freedoms; she has made a distinction between the indicators of c) autonomy and d) ability in measuring agency, seemingly trivial, but critical according to Sen's agency formulation, to give just a few examples

(Alkire, 2008, 13–16, 19–21). In my study, educational well-beings and the exercise of agency of the women are presumed to be constructed in relation to their capability sets, understood as resources and socio-cultural meanings alike. Evidently, the relational and constructed ontological understanding of the women's educational pathways implies that there is an interconnection between structure and agency, a prerequisite to be analysed empirically (cf. Kabeer 1999).[10]

To add one more concept, *empowerment* is conceived as the expansion of agency (Samman and Santos 2009; cf. Burchardt, Evans and Holder 2012), yet, as noted by Alkire (2005, 220), 'empowerment' is not a term that the CA often employs. In my study too (barring the reference to the term in Figure 2.1 in describing the process towards educational well-being and agency), the concept of 'empowerment' is not applied. Neither is agency 'measured', as has been done in the World Bank's empowerment examinations by Narayan (2002, 2005), Alsop and Heinsohn (2005) and Alsop, Bertelsen and Holland (2006), although there is a strong resemblance between their approach and how I investigate the interaction between the agency and opportunity structures by analysing the i) existence of choice ii) use of choice and iii) the achievement of choice, resonating with the four constitutive elements involved in the exercise of agency. Kabeer (1999, 45) has also drawn an analytic line between 'efficient agency' (making decisions within particular social constraints) and 'transformative agency' (decisions that impact on one's social and/or economic status), denoting that she rejects the idea of agency being equalised only with choices and decision-making. Another remark concerning the 'measurement' of agency that Kabeer (1999) points out relates to resources and assets and their presumed correlation with agency. Recalling the two Tanzanian girls who fail in their mathematics exams, and considering that they have the same amount of money to be spent on private tuition: if their agency is measured on the basis of resources, then their respective agencies would be equal; if, however, one desires to go to study mathematics and science subjects at the university, but cannot do that because she failed the exam due to family restrictions, whereas the other wants to leave school and go into acting, then, due to a diversity in their target setting and desires, and their ability to pursue the agency goals and objectives that they value and have a reason to value, their agency achievements and agency freedoms are unequal (cf. Alkire 2008). Evidently, using resources and assets as proxy indicators of agency is problematic, as is the use of functionings (alone) as indicators of well-being. For this reason, as was previously argued, the CA claims that whatever concept of advantage one chooses to consider, the information base of the judgement ought to correspond with that.

In my study, I do not aim to give *indicators* for agency, neither do I *measure* agency, let alone make *comparisons* as such between the research participants' agencies. Although I recognise the not so straightforward characteristic of the conception of agency, as denoted by Kabeer, Alkire, Narayan, and Alsop and her colleagues, instead of engaging in conceptual controversies, and along with the position taken with the methodological strategies of 'sociologising philosophy' and 'rational reconstruction', the comprehension of agency is here equivalent to the choices and decision-making processes, referring to the research participant's feeling of agency (see Welzel and Inglehart 2010), as a personal and unique experience and perception. In addition, I do examine the availability and relevance of resources, signifying factors that support converting opportunities into functionings. However, resources and assets are not presumed to correlate with the (feeling of) agency, notwithstanding their evident connection with educational well-being. Besides, resources such as schools, classroom conditions, pedagogical knowledge and skills of the teachers are exactly the kind of supply-side factors that governments, including the government of Tanzania, are investing in, presuming a correlation between them and the educational well-being of students.

The final remarks to note concern the 'measurement' of agency that seems to emerge in the literature on women's agency. First, note its inherently multi-dimensional nature, denoting different levels of agency at different stages of women's life cycles and in different spheres of life (see the Agency and Empowerment review by Samman & Santos 2009; see also Alkire 2005, 225). Gilligan (1982, 71), for example, discusses the dilemma between femininity, which involves responsibility to others, and adulthood (or masculinity), which is commonly associated with autonomy. As exemplified by Alkire (2008, 11), the same person can be fully empowered as a wife and a mother, but excluded from the labour force due to social conventions; the same person may be recently empowered to vote by a grassroots political process, but may not feel confident to travel alone. This is of importance for the participants in this research, who negotiate and seek to find ways in which to exercise agency as professionals, breadwinners and students, mothers and wives, and daughters and daughters-in-law, which all embed different and intersecting assumptions and expectations to be reconciled with their own *ideas* concerning their different life-spheres. Second, it is pivotal to comprehend agency, similarly to the concept of gender, as relational in an ontological sense, constructed in relation to others, and to the contextualising and preconditioning social structures (institutions and meanings), which in this study signifies 'Us' and 'They' and the two definite enabling environments respectively. Third, and because of the relational ontological position, agency needs to be seen as a cultural and context-specific concept (Samman and Santos 2009). The last two remarks converge with the previously discussed criticism of the status quo attached to *the idea of gender* that is neither natural/inevitable nor unchangeable (Chafetz 1988; Hacking 2000; Marshall 2008). However, these utterances are methodologically suggestive of case-by-case kinds of studies, having somewhat little evidence to inform the design of education development policies, for instance (cf. Clark 2009[11]). Yet in spite of the limitations of making a generalisation *inter alia*, it is important to acknowledge that people's agency 'is deeply informed by their own knowledge and values, and research that identifies people's varied understandings of appropriate agency may catalyse constructive public discussions that further shape how people value their own and other's agency' (Alkire 2008, 7). This is, indeed, the rationale and justification with which to obtain a fuller and more nuanced understanding of the phenomenon of girls' and women's education and schooling in the global South and Tanzania by narrating the stories of the women, and by being informed of their experiences and insights regarding educational well-being and agency.

What then is understood as constructing the educational well-being and agency of the participants in this research? *Functionings* are educational beings and doings, which the women have *achieved*, including 'access to' and 'participation in' education; learning, performing, completing basic education; access to university, and pursuing Master's degrees, to give some examples. Hence, their educational functionings are the kinds of *facts* relating to formal education, which are represented, for instance, in the national statistical data, and are examinable retrospectively in this case by asking: What have the women achieved? In my study it was more relevant to ask, however, *how* their functionings were achieved. Therefore, as has been repeatedly argued, capabilities were more accurate indicators for analysing this aspect of educational well-being.

Capabilities are understood as the women's *opportunities* to *realise* educational functionings. The presupposition is that, even though we may see an equal outcome/functioning (e.g. university degree), the women's opportunities to function might have been different. There is nothing inherently unequal about the differences, but the key issue is that differences, such as disability, age and the available amount of money, may become inequalities. Hence, the practical adaptation of the CA is to examine 'structurally given freedoms', to be comprehended as a *set of opportunities* open to the women, virtually synonymous with enabling environments, which indeed enabled them to access and participate in education, to learn and perform and to reach university level and pursue a degree – in contrast to the majority of Tanzanian women. Additionally, the analytical interest is directed at identifying the means with which to achieve educational well-being, by which I mean the resources to be converted into functionings. Thus, in this examination the research participants' capabilities are understood as i) opportunities to be educated, implying that there are at least two alternatives available, for example, to be educated in the first place or not, *and* ii) the potential of resources to be utilised in achieving functionings. There may be differences produced by gender, however, amongst people when it comes to decision-making and choosing between alternatives, and what they can actually do with the resources that they possess, for example, as was exemplified by the use of a bike, and this leads us to the concept of freedom being at the core of the research participants' notion of their agency.

Women's *agency* notions are defined to be examined on the basis of the question: Who decides? Hence, different from capabilities, based essentially on the 'structurally given freedoms', research participants' feeling of agency mainly comprises 'reasoning-based freedoms'. This in turn is suggestive of i) the freedom to set agency goals, which are the various beings and doings that the women have had a reason to value, and ii) the freedom to act accordingly. In addition to agency goals and agency freedoms, another important aspect concerning the research participants' educational well-being and agency arises on the basis of the concept of agency achievement, which refers to the particular level of education, the particular level of attainment, particular subject choices etc., which is the realisation of goals and values that they have had a reason to pursue. The key analytical distinction with the functionings is that the agency achievements (personally valued) are not necessarily the same as those that the women have actually achieved, and indeed, this probable divergence is at the core of my study. There are overlaps between well-being freedoms (capabilities) and agency freedoms, and they intertwine. The key analytical distinction between the two categories can be illustrated by asking and answering two different kinds of questions: (i) What kind of resources, opportunities and alternatives did the women have to choose to construct their educational career and (ii) Who decided what to reach for, and how, when and under what preconditions? In other words I asked: to what extent the research participants have had the freedom to exercise educational agency, which includes opportunities and alternatives (well-being freedoms), and the autonomy to make decisions and take actions (agency freedom) with regard to their educational pathways; to what extent are their beings and doings in accordance with the *idea of Tanzanian woman*; and to what extent are they aimed at becoming something else? To be more precise, I have been expressly interested in learning whose reasoning matters, and who made the decisions concerning the women's educational pathways and other areas of their life: the women, their parents and/ or relatives, their husbands and/or in-laws, no one in particular, but the system; or all of them, intertwined into a complex web of processes of becoming.

In Chapter 5, the various conversion factors that block girls' and women's educational well-being and agency were presented. In Chapters 8 and 9, the research participants' school, social and familial environments are analysed as their sets of capabilities with which to construct educational well-being, and some attention is also given to the aspect of agency. The analytical implication of the capabilities approach focusing on the notion of research participants' agency (in relation to the sets of opportunities, in other words, enabling environments or social structures) is analysed more in Chapter 10.

Notes

1 For Sen (1992, 44; also 1999) education is one of 'a relative small number of centrally important beings and doings that are crucial to well-being'. Similarly, for Nussbaum (1997, 2000, 2004) education is a basic capability *par excellence*. In the writings of both, the importance of education for 'human flourishing' and dignity is underlined, and the benefits for women's capabilities and agency are acknowledged and attested to. Nussbaum, in particular, requires the proper functioning of government to make available the basic necessary conditions, including educational institutions and arrangements, for a fully good life (in accordance with her universal list of central capabilities, in contrast to the thinking of Sen, e.g. 1993, 2004). The question of a list

in the educational context has been discussed e.g. by Terzi (2007), with regards to basic capabilities, and Walker, in the context of gender equality (2006, 2007).

2 Alkire (2008) discusses the internal pluralism of Sen's account of agency, concerning both its concept and measurement. One of the key points she makes is how agency, by definition, interrelates with goals that the person values, and thereby with the conception of the good. She also notes that, although for many people capability is related to agency, agency and well-being perspectives remain importantly distinct (Alkire 2008).

3 Bourdieu (1977), for example, has returned to the analysis of 'the field of education' time and again as in his early education research compilation 'La Reproduction' with Passeron (Bourdieu and Passeron 1977; see also the 'Bourdieu' special issue in the *British Journal of Sociology of Education 25*, no. 4, 2004. An interesting and very recent deployment of Bourdieu's thinking is the 'New Model of Social Class?' in the UK by Savage et al (2013).

4 For example, Nelly Stromquist (1989) on gender and Harry Brighouse (2000, 2002) on school choice and social justice and the rights of the children.

5 Upgrading qualifications has been identified by Tao (2013) as one of the most meaningful occupational functionings amongst the Tanzanian teachers she interviewed.

6 Walker and Unterhalter (2007, 12) have coupled Sen's notion of freedom to Dewey and Freire, to place equal value on the processes of decision-making and on the opportunities to achieve valued outcomes, emphasising that we make development and freedom 'by doing' development and freedom.

7 Children's capabilities to aspire to and consider change in four African countries including Tanzania have been recently discussed by Unterhalter (2012), and the future orientation of young Tanzanian women by Posti-Ahokas (2012); children and their future expectations in the European context are discussed e.g. by Klasen and Halleröd in Otto and Ziegler (2010).

8 On children and the capability approach (Biggeri, Ballet and Comin 2011).

9 Watts (2009; Watts and Bridges 2006) has analysed adaptive preferences in higher education, as part of the twin policy concerns of economic development and social justice in the UK, exemplified by the aspirations of young people who have rejected higher education. Walker, similarly, has discussed social justice in education and capabilities in the European context (2010). The global debates concerning universities, and the role of higher education in advancing equalities, are at the core of a very recent publication by Boni and Walker (2013).

10 See also Arnot et al. (2012a) for a discussion on the presumed linear transition from school to work and adulthood attached to the metaphoric concept of a 'pathway'; cf. Posti-Ahokas and Okkolin (2015) referring to non-linear and 'rocky pathways'.

11 Clark discusses Nussbaum's empirical evidence concerning adaptive preferences by providing only anecdotal evidence; yet these are powerful and compelling narratives of the women. See also Unterhalter's (1999) 'schooling autobiographies' of two groups of South African women.

Bibliography

Alkire, S. 2002. *Valuing Freedoms: Sen's Capability Approach and Poverty Reduction*. Oxford: Oxford University Press.

Alkire, S. 2005. Subjective Quantitative Studies of Human Agency. *Social Indicators Research* 74: 217–260.

Alkire, S. 2008. Concepts and Measures of Agency. OPHI Working Papers No. 9. Oxford: OPHI.

Alkire, S. 2010. Human Development: Definitions, Critiques, and Related Concepts. OPHI Working Paper No. 36. Oxford: OPHI.

Alsop, R., M. Bertelsen, and J. Holland. 2006. *Empowerment in Practice: From Analysis to Implementation*. Washington D.C.: World Bank.

Alsop, R., and N. Heinsohn. 2005. *Measuring Empowerment in Practice: Structuring Analysis and Framing Indicators*. Washington D.C.: World Bank.

Arnot, M., R. Jeffery, L. Casely-Hayford, and C. Noronha. 2012a. Schooling and Domestic Transitions: Shifting Gender Relations and Female Agency in Rural Ghana and India. *Comparative Education* 48, no. 2: 181–194.

Ballet, J., J.-L. Dubois, and F.-R. Mahieu. 2007. Responsibility for Each Other's Freedom. Agency as the Source of Collective Capability. *Journal of Human Development* 8, no. 2: 185–201.

Biggeri, M. 2007. Children's Valued Capabilities. In *Amartya Sen's Capability Approach and Social Justice in Education*, ed. M. Walker and E. Unterhalter, 197–214. New York: Palgrave Macmillan.

Biggeri, M., J. Ballet, and F. Comin, eds. 2011. *Children and Capability Approach*. New York: Palgrave Macmillan.

Biggeri, M., R. Libanora, S. Mariani, and L. Menchini. 2006. Children Conceptualising Their Capabilities: Results of a Survey Conducted During the First Children's World Congress on Child Labour. *Journal of Human Development and Capabilities* 7, no. 1: 59–83.

Biggeri, M., and M. Santi. 2012. The Missing Dimensions of Children's Well-Being and Well-Becoming in Education Systems: Capabilities and Philosophy for Children. *Journal of Human Development and Capabilities* 13, no. 3: 373–395.

Boni, A., and M. Walker, eds. 2013. *Human Development and Capabilities: Re-imaging the University of the Twenty-First Century*. London: Routledge.

Bourdieu, P., and J.-C. Passeron. 1977. *Reproduction in Education, Society and Culture*. London: Sage.

Brighouse, H. 2000. *School Choice and Social Justice*. Oxford: Oxford University Press.

Brighouse, H. 2002. What Rights (If Any) Do Children Have? In *The Moral and Political Status of Children*, ed. D. Archard and C. MacLeod, 31–52. Oxford: Oxford University Press.

Burchardt, T. 2009. Agency Goals, Adaptation and Capability Sets. *Journal of Human Development and Capabilities* 10, no. 1: 3–19.

Burchardt, T., M. Evans, and H. Holder. 2012. Measuring Inequality: Autonomy – The Degree of Empowerment in Decisions about One's Own Life. CASE report No.74. London: The Centre for Analysis of Social Exclusion.

Carr-Hill, R., and J. Ndalichako. 2005. *Education Sector Situation Analysis: Final Draft Report Revised*. Dar es Salaam, Tanzania: Economic and Social Research Foundation.

Chafetz, J.S. 1988. *Feminist Sociology: An Overview of Contemporary Theories*. Itasca: Peacock.

Clark, D.A. 2003. Concepts and Perceptions of Human Well-Being: Some Evidence from South Africa. *Oxford Development Studies* 31, no. 2: 173–196.

Clark, D.A. 2009. Adaptation, Poverty and Well-Being: Some Issues and Observations with Special Reference to the Capability Approach and Development Studies. *Journal of Human Development and Capabilities* 10, no. 1: 21–42.

Crocker, D.A., and I. Robeyns. 2010. Capability and Agency. In *Amartya Sen*, ed. C.W. Morris, 60–90. Cambridge, UK: Cambridge University Press.

Gadamer, H.-G. 1975. *Truth and Method.* New York: Crossroads.

Gilligan, C. 1982. *In a Different Voice.* Cambridge, MA: Harvard University Press.

Hacking, I. 2000. *The Social Construction of What?* Cambridge, MA: Harvard University Press.

Hart, C.S. 2009. Quo Vadis? The Capability Space and New Directions for the Philosophy of Educational Research. *Studies in Philosophy and Education* 28, no. 5: 391–402.

Kabeer, N. 1999. Resources, Agency, Achievement: Reflections on the Measurement of Women's Empowerment. *Development and Change* 30, no. 3: 435–464.

Kabeer, N. 2005a. *Social Exclusion: Concepts, Findings and Implications for the MDGs: Commissioned Paper.* London, UK: DFID. http://www.gsdrc.org/docum ent-library/social-exclusion-concepts-findings-and-implications-for-the-mdgs/

Kabeer, N. 2005b. *Inclusive Citizenship: Meanings and Expressions.* London: Zed Books.

Kaufman, A. 2006. Capabilities and Freedom. *Journal of Political Philosophy* 14, no. 3: 289–300.

Kelly, A. 2012. Sen and the Art of Educational Maintenance: Evidencing a Capability, as Opposed to an Effectiveness, Approach to Schooling. *Cambridge Journal of Education* 42, no. 3: 283–296.

Lehtomäki, E. 2005. Pois oppimisyhteiskunnan marginaalista?: Koulutuksen merkitys vuosina 1960–1990 opiskelleiden lapsuudestaan kuurojen ja huonokuuloisten aikuisten elämänkulussa. Jyväskylä Studies in Education, Psychology and Social Research No. 274. Jyväskylä, Finland: Jyväskylän yliopisto.

Lessmann, O. 2009. Capability and Learning to Choose. *Studies in Philosophy and Education* 28, no. 5: 449–460.

Malhotra, A., and S.R. Schuler. 2005. Women's Empowerment as a Variable in International Development. In *Measuring Empowerment: Cross-Disciplinary Perspectives,* ed. D. Narayan-Parker, 71–88. Washington D.C.: World Bank.

Marshall, B.L. 2008. Feminism and Constructionism. In *Handbook of Constructionist Research,* 687–700. New York: The Guilford Press.

Mbilinyi, M. 2000. *Gender Issues in Higher Education and their Implications for Gender Mainstreaming and Strategic Planning.* Dar es Salaam: UDSM Gender Dimensions Programme Committee.

Mbilinyi, M. and P. Mbughuni, eds. 1991. *Education in Tanzania with a Gender Perspective. Summary Report.* Dar es Salaam/Stockholm: SIDA.

McNay, L. 2000. *Gender and Agency. Reconfiguring the Subject in Feminist and Social Theory.* Cambridge: Polity Press.

McNay, L. 2003. Agency, Anticipation and Indeterminacy in Feminist Theory. *Feminist Theory* 4, no. 2: 139–148.

Moser, A. 2007. *Gender and Indicator: Overview Report.* Brighton, UK: Institute of Development Studies.

Narayan, D. 2002. *Empowerment and Poverty Reduction.* Washington D.C.: World Bank.

Narayan, D., ed. 2005. *Measuring Empowerment: Cross-Disciplinary Perspectives.* Washington D.C.: World Bank.

Nebel, M., and T. Herrera Rendon. 2006. A Hermeneutic of Amartya Sen's Concept of Capability. *International Journal of Social Economics* 33, no. 10: 710–722.

Nussbaum, M. 1997. *Cultivating Humanity: A Classical Defence of Reform in Liberal Education.* Cambridge, MA: Harvard University Press.

Nussbaum, M. 2000. *Women and Human Development: The Capabilities Approach.* Cambridge, UK: Cambridge University Press.

Nussbaum, M. 2001. Adaptive Preferences and Women's Options. *Economics and Philosophy* 17: 67–88.

Nussbaum, M. 2003. Political liberalism and respect: A response to Linda Barclay. *Nordic Journal of Philosophy* 4: 25–44.

Nussbaum, M. 2004. Women's Education: A Global Challenge. *Signs: Journal of Women and Culture in Society* 29, no. 2: 325–355.

Nussbaum. M., and A. Sen. 1993. *The Quality of Life: Studies in Development Economics.* Oxford: Oxford University Press.

Okkolin, M.-A., E. Lehtomäki, and E. Bhalalusesa. 2010. Successes and Challenges in the Education Sector Development in Tanzania: Focus on Gender and Inclusive Education. *Gender and Education* 22, no. 1: 63–71.

Otto, H.-U., and H. Ziegler, eds. 2010. *Education, Welfare and the Capabilities Approach.* Opladen: Barbara Budrich Publishers.

Posti-Ahokas, H. 2012. Empathy-Based Stories Capturing the Voice of Female Secondary School Students in Tanzania. *International Journal for Qualitative Studies in Education* 26, no. 10: 1277–1292.

Posti-Ahokas, H., and M.-A. Okkolin. 2015. Enabling and Constraining Family: Young Women Building Their Educational Paths in Tanzania. *International Journal of Community, Work and Family.* http://dx.doi.org/10.1080/13668803.2015.1047737.

Reinharz, S. 1992. *Feminist Methods in Social Research.* Oxford: Oxford University Press.

Robeyns, I. 2005. The Capability Approach: A Theoretical Survey. *Journal of Human Development* 6, no. 1: 93–114.

Robeyns, I. 2006b. The Capability Approach in Practice. *Journal of Political Philosophy* 14, no. 3: 351–376.

Robeyns, I. 2011. The Capability Approach. *Stanford Encyclopedia of Philosophy (Summer 2011 edition),* June 4, 2012.

Saito, M. 2003. Amartya Sen's Capability Approach to Education: A Critical Exploration. *Journal of Philosophy of Education* 31, no. 1: 17–33.

Samman, E., and E. Santos. 2009. *Agency and Empowerment. A Review of Concepts, Indicators and Empirical Evidence (Research in Progress 2009).* Oxford: OPHI.

Savage, M., F. Devine, N. Cunningham, M. Taylor, Y. Li, J. Hjellbrekke, B. Le Roux, S. Friedman, and A. Miles. 2013. A New Model of Social Class? Findings from the BBC's Great British Class Survey Experiment. *Sociology* 47, no. 2: 219–250.

Sen, A. 1985a. Well-being, Agency and Freedom: The Dewey Lectures 1984. *Journal of Philosophy* 82, no. 4: 169–221.

Sen, A. 1985b. *Commodities and Capabilities.* Amsterdam: New Holland.

Sen, A. 1992. *Inequality Re-examined.* Oxford: Oxford University Press.

Sen, A. 1993. Capability and Well-being. In *The Quality of Life: Studies in Development Economics,* ed. M. Nussbaum and A. Sen, 30–53. Oxford: Oxford University Press.

Sen, A. 1999. *Development as Freedom.* Oxford: Oxford University Press.

Sen, A. 2002. *Rationality and Freedom.* Cambridge, MA: Harvard University Press.

Sen, A. 2004. Capabilities, Lists, and Public Reason: Continuing the Conversation. *Feminist Economics* 10: 77–80.

Stromquist, N. 1989. Determinants of Educational Participation and Achievement of Women in the Third World: A Review of the Evidence and a Theoretical Critique. *Review of Educational Research* 59, no. 2: 143–183.

Subrahmanian, R. 2005. Gender Equality in Education: Definitions and Measurements. *International Journal of Educational Development* 25: 395–407.

Tao, S. 2013. Why Are Teachers Absent? Utilising the Capability Approach and Critical Realism to Explain Teacher Performance in Tanzania. *International Journal of Educational Development* 33: 2–14.

Terzi, L. 2007. The Capability to be Educated. In *Amartya Sen's Capability Approach and Social Justice in Education*, ed. M. Walker and E. Unterhalter, 25–43. New York: Palgrave Macmillan.

UNDP. 2001. *Learning & Information Pack: Gender Analysis.* New York: UNDP.

Unterhalter, E. 1999. The Schooling of South African Girls. In *Gender, Education and Development: Beyond Access to Empowerment*, ed. C. Heward and S. Bunwaree, 49–64. London: Zed Books.

Unterhalter, E. 2003. The Capabilities Approach and Gendered Education: an Examination of South African Complexities. *Theory and Research in Education* 1, no. 1: 7–22.

Unterhalter, E. 2007a. *Gender, Schooling and Global Social Justice.* London: Routledge.

Unterhalter, E. 2007b. Gender Equality, Education, and the Capability Approach. In *Amartya Sen's Capability Approach and Social Justice in Education*, ed. M. Walker and E. Unterhalter, 87–107. New York: Palgrave Macmillan.

Unterhalter, E. 2012. Inequalities, Capabilities and Poverty in four African Countries: Girls' Voice, Schooling, and Strategies for Institutional Change. *Cambridge Journal of Education* 42: 307–325.

Unterhalter, E., and H. Brighouse. 2007. Distribution of What for Social Justice in Education. In *Amartya Sen's Capability Approach and Social Justice in Education*, ed. M. Walker and E. Unterhalter, 68–86. New York: Palgrave Macmillan.

URT (United Republic of Tanzania). 1995. *Education and Training Policy (ETP).* Dar es Salaam, Tanzania: Ministry of Education and Culture.

URT. 2010a. *Basic Education Statistics in Tanzania 2006–2010.* Dar es Salaam, Tanzania: Ministry of Education and Vocational Training.

Vaughan, R. 2007. Measuring Capabilities. In *Amartya Sen's Capability Approach and Social Justice in Education*, ed. M. Walker and E. Unterhalter, 109–130. New York: Palgrave Macmillan.

Walker, M. 2006. Towards a Capability-Based Theory of Social Justice for Education Policy-Making. *Journal of Education Policy* 21, no. 2: 163–185.

Walker, M. 2007. Selecting Capabilities for Gender Equality in Education. In *Amartya Sen's Capability Approach and Social Justice in Education*, ed. M. Walker and E. Unterhalter, 177–195. New York: Palgrave Macmillan.

Walker, M. 2010. Capabilities and Social Justice in Education. In *Education, Welfare and the Capabilities Approach*, H.-U. Otto and H. Ziegler, 155–179. Opladen: Barbara Budrich Publishers.

Walker, M., and E. Unterhalter. 2007. The Capability Approach. In *Amartya Sen's Capability Approach and Social Justice in Education*, 1–18. New York: Palgrave Macmillan.

Watts, M. 2009. Sen and the Art of Motorcycle Maintenance: Adaptive Preferences and Higher Education. *Studies in Philosophy and Education* 28, no. 5: 425–436.

Watts, M., and D. Bridges. 2006. Enhancing Students' Capabilities? UK Higher Education and the Widening Participation Agenda. In *Transforming Unjust*

Structures, ed. S. Deneulin, M. Nebel, and N. Sagovsky, 143–160. Dordrecht, Netherlands: Springer.

World Bank. 2013. *On Norms and Agency, Conversations about Gender Equality with Women and Men in 20 Countries.* Washington DC: World Bank.

Welzel, C. and R. Inglehart. 2010. Agency, Values, and Well-Being: A Human Development Model. *Social Indicators Research* 97: 43–63.

Part III
Women's Many Stories

8 Educational Well-being in Schools

Leyla's Story

As we learned from the previous research, the factors related to the school and learning environment (long and busy school days, distance from and travelling to school, the issues of food, classroom conditions, fellow students, school practices, teachers' attitudes and pedagogical skills, curricula, etc.), affect not only female access to education, but the girls' and women's well-being at school as well. To discuss these factors, and to learn *from* Leyla's narrative, in line with the key idea of the vcr-method, before I start talking *about*, and analysing her and other women's experiences and insights, I give room for her first to speak about her primary and secondary school experiences and insights (Connell 2007; Reinharz 1992).

My primary school days were **very busy**. I woke up at **5am** in the morning and made **some tea** for me and for my brother and sisters. After that I went to school **on foot** with my **girlfriends** who were in our neighbourhood. My school was about **five kilometres** away [...]. We had to run all the way to school because if you got there late, you were **caned**. At seven o'clock in the morning, the bell rang for us to start **cleaning** the school compound. Every student had a plot to sweep or pick up papers. Others had flower gardens to water. Everyone had a day when he or she was supposed to clean the classroom and the toilets. Failure to do your duty resulted in punishment.

At half past seven, the bell rang again. All students stood in lines and teachers went around **inspecting** our uniforms, our teeth, fingernails, hair etc. and if found dirty, you were punished. While the teachers were inspecting us, the school band was playing some music. After inspection, we all sang the national anthem led by the school band. Then the head teacher gave some announcements: as an example, those who had not paid **fees for food** or had a **torn uniform** were called in front of the whole school and were told to go home until when they could pay; it was a time also when the teacher on duty called out the names of latecomers for that morning to get the caning. If they were a big number, they were given a plot in the garden to dig when others left for home in the evening.

In my class, we were about **35 pupils (boys and girls)**. The teachers forced us to mix with the **boys** when sitting. Then classes started. But when the teacher went out of the class for some reason, we re-arranged ourselves and boys sat alone separate from the **girls**. When the teacher caught us doing that, we were often scolded or even caned. At school, **I enjoyed learning English** the most. The teacher was good and encouraged us to learn. He also gave me incentives (like food from his lunch). He was like a grandfather to me. He hated it when we (girls) became naive. He told us about his life experiences as a teacher in distant schools. I personally regarded him as a 'hero' of some kind. I didn't want to let him down by failing his subject. At school, I didn't like much about the caning of the pupils, especially in the classroom. I could avoid getting late to school but I could not avoid making mistakes in my studies. Caning was the order of the day, e.g. in **maths**, learning the multiplication tables by heart was difficult and the teacher showed no mercy. We had maths the first thing in the morning, and it was very cold. We were caned until our fingers turned blue with numbness. But **we managed to learn somehow**. (Girls were beaten on their palms and boys on their buttocks.) At noon, we had **lunch break**, and then we went back to classes around 1.30 p.m. We studied until 3.30 p.m. when the bell rang and we all lined up outside again, some more announcements from teachers and canning of wrong-doers, then at 4.00 p.m. we were allowed to go home after singing a farewell song.

After seven years in the primary school, we did the fearful 'final' **examinations**. The results came out in December during Christmas holidays. The rumours had circulated that I was among the two students who had been **selected to go for further studies**. I have **never felt happier and proud of myself as that day**. My father called me in his sitting room and gave me the news. I rushed out into the kitchen where I was cooking the supper and threw some water on the fire. My young sisters and brother didn't understand why I did that. 'Bye-bye firewood and smoke in my eyes' I told them, "bye-bye caning!' I was going to girls' secondary school. They all rushed out and shouted with joy. My elder brother was at home during the holidays. He told me not to accept inferior **quality of clothes** and suitcase to go to secondary school because I would be laughed at by other students. He told me secretly to go with our father to town so that I could refuse such clothes and suitcase. Unfortunately, the next day my father came home with all the needed things already bought. **No choice!** But I was not worried. Passing the exam was more important.

School was to me a very **beautiful school**, run by missionaries. **All girls, very few**; we knew everybody in the school. We had very **nice teachers**. A school day started at **6.00 a.m.** when the big bell rang to wake us up. We took our baths in a large **bathroom**, naked, everybody, was everybody and we were not supposed to be shy. It was difficult at first, but later I got used to it. We cleaned our dormitories and toilets in turn. Outside, there were workers employed to cut the lawn and to water the flowers. At 7.00

a.m., the breakfast bell rang. We went to the dining hall and had porridge on some days and tea with pancakes on others. At 7.30 a.m. the bell went again for us to go for prayers in the chapel. One of us or a teacher would lead us to sing a hymn and pray. A student would also play a hymn on the piano. **It was wonderful!**

At **8.00 a.m.,** the bell for classes rang and we all went to our separate classes. At Form I and II we learned **both arts and science** subjects, but at Form III we could choose, under the advice of our teachers, to join either arts or science streams depending on the ability. **The teachers were so good. No more caning!** Even maths was no longer a terror to me. At 10.30 a.m., it was 30 minutes **tea-break**. We rushed to the dining hall for a cup of tea. No cookies! Then back to class until **2.00 afternoon. It was the end of classes**. We went for **lunch**. After lunch, personal cleanliness & reading for pleasure or resting. At 6.00 p.m., we had **supper**. Then at 7.30 p.m. we had **preparation** in the classrooms until 9.30. Then prayers, then at **10.00 lights were off**. Everybody slept.

The issues highlighted (bolded) in the story of Leyla will be examined in detail in the following sections. There are three reasons to give such a detailed and close look at the school environment (including school infrastructures, human relations at the schools and teaching and learning experiences). The first reason is that this is the area where the strategic priorities and practices of the educational policy reforms are primarily directed. The second reason is that the school environment, including the infrastructure, school and classroom practices, and human relations with other students and teachers, may be comprehended as the kinds of concrete and real 'opportunities' and presence of 'options', what Sen (1992) means by capabilities. The third is because I am particularly interested in giving voice to policy implementation, by which I mean school-related practices, experienced by individual students, in comparison to the ideas that 'exist only formally in policy agenda' (ibid.) To begin with, I will provide an overview, however, of the issue of poverty and the costs of schooling at the time of the research participants' primary and secondary education. Why? Because most of the research on education, gender and development report the direct costs of schooling to be the major reason for parents either not educating or removing children, particularly girls, from school, which is not surprising given the prevailing poverty in sub-Saharan Africa (Odaga and Heneveld 1995; see also Colclough, Rose and Tembon 2000; Colclough et al. 2003; King, Palmer and Hayman 2005; UNICEF 2012).

Costs of Schooling

Schooling costs such as tuition and other costs such as fees for registration and admission, examinations, boarding, the school building fund, parent and teacher association fees, uniforms, the provision of furniture, extra tutorials, and transport, are beyond the means and resources of many families, and reportedly affect girls' education and schooling more than that of boys. As an example, for modesty reasons, girls are less likely to go to school in torn or ill-fitting uniforms, as we will learn from Leyla. In addition, for safety reasons, parents tend to spend more money on transport. For the most part, the literature on education, gender and development presented earlier, examining the linkages between poverty and educational decision-making, suggests that girls from better-off homes, who live in urban areas, are more likely to be enrolled and remain in school longer than those from poorer homes and rural areas. This is also echoed to some extent, in the narratives of the research participants.

For their primary education, all of the research participants attended government schools, seven of them in an urban environment and three in village schools. Although the education was free, the costs of schooling still included 'uniforms, shoes, books and exercise books, pens, writing papers; everything the parents had to buy', listed Wema; this was in addition to travel costs for some of the women. Genefa, who along with her three siblings was dependent on their older sister, described how the free education encouraged them to go to school, and luckily, they all passed Std VII examinations, which enabled them to continue further. As explained earlier, transition to upper grades in primary school and to reach secondary education depends on the results in the national examinations.

> If we would have failed, it wouldn't have been possible for us to continue because my sister was the only one who provided assistance for us. At the same time, we were doing small business to earn money for uniforms and exercise books.
>
> (Genefa, 40)

Even though there were no school fees in primary school, the other schooling costs turned out to be critical and even prohibited schooling for Leyla:

> Primary school was free, but not very free, because they (parents) had to buy the uniform and they had to pay lunch. We had to have lunch in school. It was hard on them, I could see, feeding us (six siblings). [...] Sometimes I stayed home because, for example, my father hadn't paid the fees for lunch, or if the uniform was torn, and my father couldn't afford at that time a new one. Sometimes my father wrote some notes to the headmaster and I was allowed to be in class.
>
> (Leyla, 53)

The prohibitive versus supportive impact of the uniforms for school attendance is debated both in public discussion and in the research literature; the participants in this research expressed, in unison, the positive and nice memories about the school uniforms and clothing. Despite Leyla's experiences with a 'torn uniform', she talked about 'Independence shoes' when I asked her to recall her nicest school memories. As the oldest of the research participants, she started Std 1 in 1961, the year of Tanzanian independence. She remembered that she was allowed to wear shoes for the first time, 'green Independence shoes', at school. Similarly, Amisa and Wema discussed the joy and excitement of starting their schooling, and how proud they were to wear brand new school uniforms.

Regarding secondary education, different educational paths were followed by the research participants: five of them were selected to join government secondary schools, where education was free and other costs relatively low; two attended private, missionary girls' schools, and three joined both private and government schools. These women had all performed extremely well in their education and benefitted from the strict examination-based advancement. Genefa proudly described how she had been selected to attend a very famous school: 'at that time the best girls' school in the country'; similarly Tumaini had been selected to join Form V in a special school as a result of her excellent performance. Hence, most of the women had either been selected to join government secondary schools or were awarded grants to private ones. If they were not successful enough to be enrolled into free public education or be awarded grants, all the women had sought financial assistance from their (extended) families. On top of that, all the women and all their families had generated additional income for education, for example by growing and selling agricultural products, baking bread and selling cattle (Posti-Ahokas and Okkolin 2015). Genefa, who among this group of women was probably the least supported by her family, describes the consequences of the poverty of her family for her secondary school attendance:

> When I was in secondary school, I spent the most time in doing domestic activities. [...] Sometimes I spent most of the time in finding food, because at that time, the country was in economic crisis. There was a shortage of food, thus I had to look for it. I spent a lot of time in long queues.
>
> (Genefa, 40)

As summarised in Chapter 2, most of the research participants described their childhood families to be 'normal middle-class'. To exemplify what that means, Tumaini (who was enrolled in government school) recalled being given 3,000 TSH[1] pocket money for the whole term of six months, to buy some snacks, sugar and personal utensils. In comparison, her friend, whose father worked in a bank, was given 10,000 TSH. 'It was kind of a difficult life, but since we were given food at the school, the money can be used for small things', she explained. She also remembered, 'The kind of clothes I was putting on was not very nice, just the normal ones compared to those whose fathers would give them more money.' (In comparison, a pair of basic shoes would cost approximately 500 TSH.) Wema's mother, who was taking care of the children by herself after the divorce, without any support from her ex-husband, was investing everything she could afford to ensure her children's educational advancement. As an example, before primary school exams she used to hire a teacher to give private evening tuition at their house.

What needs to be acknowledged here is the fact that, regardless of the rather smooth transitions from one level to another, all the research participants emphasised their parents and families' financial constraints in pursuing education for their children. Apparently, the economic conditions of the families had a direct bearing on the women's educational well-being; indeed, all of them discussed poverty and the costs of schooling during our conversations. But then, once having 'access to' and 'participating in' education, what kinds of sets of capabilities were open to them to realise educational functionings that they had a reason to value regarding the concrete physical school environment?

Physical School Environment

In this section, I focus on the physical environment and infrastructure of the schools, and address issues such as distance, meals, water, sanitary facilities and classroom conditions (e.g. teacher-to-pupil ratio (TPR), desks, chairs and learning materials, etc.). In other words, in the following sections, I examine the 'external environmental' factors influencing female education as identified and documented by Odaga and Heneveld (1995), Brock and Cammish (1997), Colclough et al. (2003), and acknowledged and tackled by the Government of Tanzania (e.g. URT 2001, 2004, 2005).

Distance to School (with empty stomach and without water)

As indicated in the review study by Odaga and Heneveld (1995), a long distance to school is a deterrent to girls' participation and achievement in schools (also Brock and Cammish 1997; Colclough, Rose and Tembon 2000; Colclough et al. 2003). As discussed earlier, the question of 'distance' is considered two-dimensional: one dimension concerns the distance itself and the energy the children have to expend to cover it (often with empty stomachs), while the other relates to (sexual) harassment (e.g. Unterhalter 2003b). The research participants recounted a few incidents regarding the latter dimension of the issue of distance, to which I will return in the chapter that concerns human relations at the schools; the former, which also recalls Leyla's story, will be addressed next in more detail.

All the participants in this research used to walk to their primary school, either because the school was not far away, or because there was no car or any other transport available. Amisa, Genefa, Naomi, Rehema, Tumaini and Wema, who attended primary schools in Dar es Salaam, and Rabia, in the village, said that their way to school did not worsen their attendance because they did not live far from the school. However, according to Rabia, the distance to school caused several dropouts, because it was very far from many other people's houses. This means, in practice, that the children in many families used to walk about three hours to the school and then three hours back. As we learned from Leyla, her school was five kilometres away, which she defines as 'nearby'. For this reason, she used to wake up at 5 a.m., and after morning duties and some tea, she ran to school. During our discussions, she described the area where she was living as 'somehow mountains', so:

> [...] when you are going to school, you had to head up! There was lot of running until you feel some smell of blood on your chest. Because if you get late, then you get punished. You were caned.
>
> (Leyla, 53)

Amana and Wema told similar stories:

> At village, we used to run, because it was somehow far and if you got to school late, you are punished. The reporting time was 7.15 a.m.
>
> (Amana, 36)

> In primary school, we would go at 7 o'clock. You have to be there exactly at 7 a.m. The head student will sit on the gate when children are coming in, and he will write all the names of the children who are coming late, which will get punished at the end of the day.
>
> (Wema, 32)

Later, Amana moved to Dar es Salaam and went to school with a dala-dala (a local minibus), which she experienced as 'much more terrible' than in the village, not to mention that she actually needed to take two buses to reach the school:

> The conductors are very harsh to students. There are negative attitudes against children and students, because nowadays we are paying 50 TSH and others 200²or 250 TSH, depending on the route. That's the problem, I think. It's somehow not safe to children, especially girls. (...) Anything bad can happen.
>
> (Amana, 36)

Rehema travelled to school using public transport, despite her physical impairment. She did not report the travelling as such being a problem but because of the distance, the school-days turned out to be extremely long:

> My school was somehow far from home. I had to take a bus in order to reach a school. I had to wake up early in the morning, always except for weekends and during holidays. Five o'clock I was out of the bed for

preparation before going to school. I had to wash utensils or clean toilets before leaving home.

(Rehema, 30)

Likewise, Genefa also woke up very early in the morning, at about 5 a.m, to start the journey to school at 6 a.m. Seemingly, on school days, most of the women rose pretty much with the sun; many of them also described how they reached home around 5 or 6 p.m., when it was already dark. This is not, however, only because of the long school-days, which could finish as late as 4 p.m., but more about them as children just playing with their friends and schoolmates 'on the road or nearby forests and other places', as explained by Wema. Some of the women used to walk to and from the school with their sisters and brothers, describing, as Hanifa did, how that in particular made the journey safe. Leyla also told how they went to school, safely, in groups: "girls alone and boys alone'.

The question of the 'distance' with regard to the 'empty stomach' was pointed out as a problematic consideration in the compilation by Odaga and Heneveld (1995), the issue being tackled today, for instance, by NEPAD (The New Partnership for Africa's Development) with WFP and other partners in nine African countries (Madamombe 2007[3]). Food and school meals were noted also by the research participants in our conversations. Only four of these women (Amisa, Rabia, Tumaini and Leyla, who all attended either village schools or a 'village kind of a school', as Tumaini described the 'education centre' that she attended) said that they ate something during their primary school days. 'Food, yes, not that good', recalled Amisa. From Leyla's story, we learned that considering that the parents had paid for it, lunch was provided at the school. Tumaini noted also how her mother, working as a teacher in the same school where she was enrolled, provided 'some small snack, maybe banana', to her. Otherwise:

Around 10, we used to have a recess, when children with some money could get around and buy something small and for those who didn't have, you just be around trying to get some from others. [...] Then we get back to classes up to around 2 p.m. (...).

(Tumaini, 33)

Amana and Genefa explained that they did not have lunch at school but went back home for lunch. There was also a lunch break time at Wema's school:

(...) you can go home, but it depends how far your home is. So, most of my friends, they didn't go home and even I didn't [...] When I was in Std VI, a little bit older, I used to run home, but the problem was that when you run home you have to cook, because my mum was at work, and eat, and then run to school.

The issue of food in primary schools is somehow summarised by Wema, as she continues:

> At 10 o'clock, we had break time, which was mainly just for playing. I can't remember if there was any food unless if you came with food from home and put it in the bag; still it was so bad because other children can take it away from you. [...] Basically, most children did not, we didn't eat during school hour time.
>
> (Wema, 32)

Only Naomi, who was enrolled in missionary boarding school from Std V onwards, reported having a tea break in the morning, and lunch and supper during the school day. During the half-hour tea break, either tea and makande (maize and bean stew) or porridge and bread were provided. After waking up at 5.30 a.m., they made the beds, bathed and dressed, and then there were ten minutes of prayer at the assembly grounds and a half-hour mass at the church. Then, 'the charges' swept the road (cleaning, hoeing, etc.), then inspection, and their first classes. In the dining room, she along with the other girls used to take their plates (everybody had a plate with the number underneath), spoons (to be put in their pockets), and Blue Bands (brand of margarine) of their own from their lockers, to be seated at tables according to numbers. After classes, until 4 o'clock, there were two hours of work at the school premises and a short rest. Supper was served at seven in the evening. Evidently, all the research participants had experienced long work-loaded schooldays, far beyond just attending classes and learning, only 'with empty stomach', but none of the women complained for themselves, as one might expect, about being hungry. Only Rabia pondered for a moment, how 'many people, because they don't eat properly, [...] they cannot concentrate on how to read and how to do anything (…).'

Apart from not *being nourished* (defined as one of the basic human functionings in the capability literature); Rabia reported that in their remote village school, there was no water:

> We didn't have water, so, every day when you go to school, you have to get a bucket of water, so that in the school they can cook the lunch and tea in the morning. [...] If you want to drink, you had to go nearby to the houses and ask for water. There was a water problem because there were only wells and you have to fetch the water far away.
>
> (Rabia, 30)

Amisa, in a big government primary school in Dar es Salaam, considered the water facilities in her school to be okay, in comparison to the situation in Third World countries, and the water to be 'safe, I think'. Similarly, Genefa appraised the water and sanitary facilities to be 'okay; very nice' in her first school with few pupils. On the other hand, Hanifa was slightly concerned about the purity of the water, since they were not given any information on it:

> If you feel thirst, you rush to a tap, you open it, you put your mouth there and you drink; there's no glasses, no cups, no whatever, and it was not as clean as maybe boiled water. It was direct from the tap.
>
> (Hanifa, 33)

In comparison with the women's experiences in secondary education, only Tumaini and Leyla, who attended government secondary schools, explained the availability and quality of water to be problematic.

> There was water shortage. We used to fetch some water for most of the time. Sometimes, that could also reduce time which could be used for reading, because after class, we went back to the dormitory, took our buckets and fetched water for cooking, cleaning etc.
>
> (Tumaini, 33)

Otherwise, because they were enrolled in single-sex boarding schools, the contrast between the learning environment in primary and secondary education concerning the infrastructure and facilities was rather dramatic for most of the research participants. As we learned from Leyla's story, the uppermost feeling when attending such 'a nice school; everything looked so beautiful', was just pure joy and happiness. Most of the women did not explicitly grumble about the distance to school, (lack of) meals, poor water and sanitary facilities, mainly because it was considered to be 'normal' or 'as usual'. Still, most of the women talked about how their secondary schools and secondary-school experiences 'were so different from the primary schools' owing to the studying and learning environment, on one hand (to be elaborated later), and to the physical environment and facilities, on the other.

'I Will Look for the Bush Instead of Going to Pit Latrine'

From a gender perspective, poverty and the costs of schooling are connected also to the sanitary facilities of the schools. To start with, the cost and/or lack of sanitary protection and underclothes for girls who have reached puberty may be the reason for not attending classes. Similarly, inadequate sanitary facilities at schools (no separate toilets for girls and boys) may have the same result. To continue, to be absent from schoolwork and classes for a week a month can undermine girls' confidence to return, and thus contribute to early dropout (e.g. Odaga and Heneveld 1995). Finally, the impact of not, for example, attending science and mathematics classes for a week, on a monthly basis, may be devastating for girls' performance and learning results in those subjects.[4] This is a concern to which the Forum for African Women Educationalists, for example, has paid special attention (e.g. FAWE 2001, 2003). In studying girls' experiences of menstruation and schooling in northern Tanzania, Sommer (2010) found that pubescent girls confront numerous challenges, such as inadequate guidance, lack of proper sanitary facilities and supplies, and gender sensitivity issues, to managing their menses in the school environment. Within the recent education sector reforms in Tanzania, among many other concerns, the necessity for sufficient sanitary facilities is explicitly noted (URT 2001, Carr-Hill and Ndalichako 2005). In Sommer's (2010) research, the girls had pragmatic and realistic recommendations for how to improve school environments, which she (ibid.) recommends being incorporated as effective methods for improving girls' academic experiences and their healthy transition to womanhood. In this study, particularly Amisa and Rabia criticised the lack of basic amenities, because 'They (male decision makers) cannot see all those small things, yah; those issues relating to women.'

Except Naomi, who completed her primary education at a girls' school, all the research participants were enrolled in co-educational schools. Basically, all the schools had separate toilets for girls and boys, although in some cases, the facilities were used in turn. For this reason, none of the women described the sanitary facilities as being peculiarly problematic or of gendered concern as such. However, many of the women mentioned that there were too few toilets in comparison to the number of students, and the quality of the facilities was considered really poor; specifically, the pit latrines were mentioned, and explicitly portrayed by Wema:

> I remember it was such a scary thing for me (...) that there is just this pit between one stone and another stone, and when you go, you have to put one leg here and another here. I don't remember if there was a child who had fallen, but I was like: I am not going there. I will look for the bush close. So, I don't remember going there so much.
>
> (Wema, 32)

As discussed earlier, most of the women attended boarding schools for their secondary education. In principle, compared to their primary schools, the physical environment was said to be considerably better: as a funny incident, Wema mentioned how she 'got used to the pit, and the stones were covered with cement'. Still, regarding the sanitary facilities from Leyla's story, we recall her being shy and modest in taking a bath with other girls. Similarly, Rehema talked about being embarrassed to share the toilet facilities. On the other hand, Rehema explained that, mainly because of her disability, she had found the school infrastructure somehow difficult: because the classes were too far from the dormitories; there were stairs to be climbed; and there was no one she could ask to go a long way to fetch the water for her. 'It was a burden to me' but 'I am not ashamed' (to ask for assistance), she concluded.

The achievements and challenges of primary education reform within the governmental undertakings concerning the improvement of learning environments, including water facilities and latrines, were discussed earlier. According to the review by HakiElimu (2005), the water situation was reported to be critical, as was the situation with desks. Conversely, parents were reported to be happy with the improved pupil-to-book ratio and the availability of other learning materials (URT 2003). In the next section, the research participants will go back to their experiences concerning learning materials and classroom conditions *inter alia*.

Classroom Conditions

The poverty of the schools is apparent in the lack of classrooms, equipment and learning materials in Tanzania, as well as in many other sub-Saharan African (SSA) countries (Odaga and Heneveld 1995). As previously discussed, the low motivation level of the teachers and the poor quality of teaching is often connected with the inadequate school environments (Tao 2013; cf. Towse et al. 2002). Within the recent education sector reforms in Tanzania, substantial school construction activities have taken place and construction of new classrooms has proven to be a major success of the Primary Education Development Plan I (PEDP 1). In addition to construction, the importance of having an adequate supply of teaching and learning resources is also outlined in PEDP I. As a matter of fact, the availability of textbooks to students was identified as a key determinant of student performance in the review by Odaga and Heneveld (1995). The influence of 'classroom tools, rules and, pedagogy' has also been analysed by Fuller and Clarke (1994). They suggest putting forward 'a culturally situated model of school effectiveness' that bridges the dichotomy between 'policy mechanics' and 'classroom culturalists', which place emphasis on school inputs (including pedagogical practices and instructional materials) and socialisation (including classroom norms and rules regarding authority, language, orientation towards achievement, conceptions of merit, status, etc.) and culturally constructed meanings (to sustain the process of socialisation) respectively. Obviously, the classroom conditions embed a parcel of opportunities and means to function and exercise agency, which are substantially intertwined.

As we remember from Leyla's story, she started her primary education in a small village school where there were about 35 students in her class. When I conversed with her, she liked to note, first, that at that time the schools were very few (not in every village as they are now), and second, that they 'were quite a handful in the classrooms' because all the children from the village went to the same school (not, however, as (over)crowded as they are now). The classrooms where Genefa she started had nearly the same number of pupils, 'we were very few' she said, but once she moved to Dar es Salaam, each class had over 50 students. She recalled that in Std VII, there were 240 students in four streams, which is pretty much the same size as Amana's reminiscences of the school from which she graduated.

Apart from Naomi, who was enrolled in a superbly equipped primary school, all of the women shared desks with other students; the number of pupils around one desk varied from three to ten students. Amana told how they sometimes sat on the floor in front of other students who had come early and reserved all the desks and seats. Similarly, Wema described sitting on the floor and hating that, not to mention the fights for the chairs:

> I hated this moment of going to school [...] At one time I even asked my parents if they could make me my own chair so that in the morning I could carry it to the school. Because it was so bad: you have this nice clean skirt, uniform, and then you go and sit on the floor; then when you go home you are totally dirty. That was worse. [...] the strong ones got the chance to sit on chair.
>
> (Wema, 32)

Earlier, Leyla mentioned being seated between boys: how she and her friends rearranged themselves and sat separate from the boys, and how they were afterwards punished for doing so. As for Hanifa, she would have preferred to stay with her twin sister in the same class and at the same desk: 'I don't know if we were afraid of what, however, it created some sort of a problem.' From the teacher's perspective, this kind of seating arrangement is argued mainly to calm noisy and harassing male students. However, from the girls' and women's perspective, this is not necessarily to advance their studying and learning. Therefore, in contrast to the disciplinary actions with Leyla, Hanifa's teacher later agreed to them being seated together with a third student. Rehema may summarise the women's experiences regarding classroom facilities:

> We shared in primary school and in O-level: we used to share desks, books and other facilities. Sometimes, when you go in your class, you find that there are no chairs and you have to go somewhere to find it and then come back to your class. It was like that. But we coped!

Regarding books, Amisa, who was enrolled in a big government school in Dar es Salaam, explained how they used to buy their own. Tumaini also explained how she used to borrow or buy her own books. She also mentioned that, because her mother was a teacher and bought books at home, she had access to books and could read whenever she wanted. In contrast, Rabia, who attended primary school in a remote village, laughed at how, 'There was only one book, which we were all supposed to read, so, you just memorise the book.' Wema, for her part, 'didn't remember seeing any books in primary school'. She mentioned that in her school there was a library but that they never actually used it. She also told that her mother used to buy exercise books for her and her siblings, which they kept at home, and never took to school. In direct contrast, according to most of the women's experiences, the books were used and stored at the schools: we had no books, they stayed at the school and were shared; even nine students would share one book, as exemplified by Hanifa:

We did not have books of our own, we shared, not only the three people sitting on one desk but even those at the front desk and those at the back desk, all of us, nine people, we shared one book. So, the book is here in the middle, these ones stand up so that they can read and those ones at the front they turn back.

'And, of course, we were not allowed to take those books to our homes', Hanifa concluded. In most of the women's cases, they could take the exercise books home, although Leyla said that those were not allowed home either, at least not until they reached Std VI and VII. As far as books and other learning materials were concerned, only one of the women, Naomi, described having had good classroom conditions. She reminded me that she was already attending good private schools at the primary level of her education:

Okay, as I told you earlier, these were the best schools, mission schools, packed up by the missionaries, so we had enough of everything: everyone could have her own novel and grammar book in English, we had laboratories, chemistry lab, physics lab, a visual room for geography. So, they were good schools.

As with the physical environment, the difference between the learning environments in the primary and secondary levels of education was remarkable for all of the women, except Tumaini. She described how it was difficult during secondary education, because now she had to buy the books, borrow them or read only the notes – in comparison to her primary education, when her mother used to bring the materials home with her. Rabia stated that, although they did not have books of their own, there was a very big library at the school: 'all kinds of books, newspapers and everything, I think that school was good'. In the same way, Hanifa had enjoyed having 'many books'; no matter that they did not have ones of their own and were still sharing. In addition, the secondary school where Leyla started had a nice environment and good facilities: 'study materials were available, some books we were sharing but some you got your own. They issued one copy and you kept it until the end of the term.' Otherwise, practically all of the women described classroom conditions and the learning environments as a whole to be better. Leyla evoked her 'rosy' experiences:

'For the first time, we went away from home and we used to like it because at home there was a lot of work to do, and we felt now we can become nice, the skin can be softer. [...] Everybody had a desk and chair of her own. We had green lawns with roses, nice houses, we were not many in the dormitories, just four, and we had separate lockers, which you could lock if you want. [...] I still hold very nice memories of my school.

Then, she was transferred to another school: 'changing from such a cosy nice school to this school, which was rough, and ah, it was such a shock'. Leyla and Rehema both compared their school experiences from O-level to A-level, and how things were actually impoverished, or as Leyla put it: 'Things were very rough, everything went wrong.' They referred, at first, to such a pivotal issue as the climate: 'It was so cold'. Rehema compared her A-level experiences in northern Tanzania to her O-level studies in Dar es Salaam: 'of course it affected my studies [...] I was going to freeze'. Leyla, on the other hand, described how she had been used to cool weather and how their teacher used to give them a 'sunshine holiday': 'We learned English outside so we could warm ourselves in the sun. It was nice. We learned!' Then, for her A-level studies, she moved to another region. She told how, for the first time, she went to a mosquito area and got malaria infections. Because of being severely ill, she did not do well in her examinations, and that is one reason why she did not perform well enough to enter university at that time. According to Leyla, another reason for failing completely was the fact that she could not understand the teachers (because of their language; the teachers were Asian by origin). But before addressing the research participants' experiences and insights concerning such a pivotal issue as the teachers, and their attitudes and peda-gogical skills, I direct the observation to human relations in schools by focusing the analysis on the women's relations with their fellow students.

Human Relations

As defined earlier, in my study the school environment is comprehended as a *fact*, but what is understood to be constructed is the way the girls and women operate, and are capable of operating in the classrooms and schools at large. In other words, as a social construction, the school environment pre-conditions and opens up a set of opportunities for girls and women's *beings and doings*, manifested in human behaviour and relations (cf. Fuller and Clarke 1994). This section pays attention to women's experiences and insights regarding their relations with other students.

Harassment and Pregnancies

As discussed in the infrastructure section, there are two ways in which the long distance to school is a deterrent to girls' participation and achievement in school. The first, which concerns the length of the distance without food and/ or water, was discussed earlier; the second, which considers the apprehension of the parents about the sexual safety of their daughters, is examined here. Indeed, this anxiety is quite justified, given the evidence presented in Chapter 2.

During our discussions, some of the research participants touched on the harassment by and offensive behaviour of the male pupils, in the classrooms and on the way to and from school. Amana and Genefa both told how they hated and feared some of the boys in their school, who used to cane the other students. Amana also recalled how some of the boys threatened to beat the girls on their way home; as a result, the girls left school before the school day finished and did not come back for the whole week – or in the worst case, at all. Amana occasionally went back to the school to avoid being chased by the boys, and was then accompanied home by the teacher. She said that this kind of harassment was somehow common at that time; but nowadays, 'There are no bushes and forests on the way to school because in the villages there are many schools: each village has one or two schools.' Wema experienced one incident when she was almost attacked by a group of boys when she was walking home from tuition and they had finished so late it was already dark:

> "Then my mother had to go and make some noise with their parents; all that fighting with their parents. Then I had to stop the tuition in the evening and had the teacher coming to the house, sometimes. But it was even more expensive, but for a while, at least when I was about to do my primary school exams, the teacher was coming."
>
> (Wema, 32)

In the same way, Tumaini felt quite unsafe to return alone from the evening classes or discussion groups from the university: 'I always pray that nothing bad happens to me on my way back from those studies', which may last until 9 p.m.

Leyla had experienced some insulting teasing during her classes but interestingly, she remembered one occasion when another girl stood up for her saying: 'If he touches you, we are going to show him; we'll teach him a lesson.' Furthermore, she also had her older brother at the same school defending and protecting her, which, she explained, she needed, because of her tiny physical size. That was, as a matter of fact, one the main reasons she was teased. Both of Leyla's secondary schools were girls' schools, but her A-level studies were at a government school, and according to her: 'it was different':

> "There were no boys to push me or threaten me with a fight (…) but the girls were harsh and rough, you know they could jump on the counter and grab the bread, and they could eat all the food and you miss it; it was a kind of a shock to us who came from elsewhere."
>
> (Leyla, 53)

Thus, not all of the teasing and harassment that was seen and experienced by the research participants were caused by the male pupils, and not all of it was sexual in nature. However, some of it was, and it resulted in unwanted pregnancies, not for the research participants, but for some of their fellow students in primary and secondary education.

> "Couple of my friends got pregnant when I was in primary school. Yes, I was really like in Std VI and I was like: Oh my God!"
>
> (Wema, 32)

The expression used by Wema was exactly the same as what I was thinking during our discussions. Indeed, many times, I had to make sure that I had understood correctly and that we were really talking about girls enrolled in primary education. But yes, in fact, all of the women that I interviewed had known some girls in their age group, between 11 and 12, during their primary education, who got pregnant and dropped out of their schooling. None of them returned, at least not to the same school. However, some of them were later re-registered in another district and school, and some of the women recalled a few relatively privileged schoolmates who were enrolled into private schools after giving birth. In comparison to the particular time that the women were referring to, in principle, today there is a policy and guidelines in Tanzania for re-entry for girls and women after giving birth. However, in practice, very few re-enter, at least the same school. Rehema considered her primary school to be 'very low in standard, so most who dropped out were girls because of early pregnancy'. Still, apart from knowing schoolmates who got pregnant during the last years of their primary education, the research participants considered that all in all, it was more boys than girls in their primary schools who actually dropped out. As an example, Rehema pointed out how half the number of pupils during her primary education dropped out. When she started, there were about 196 pupils in Std I, but only 87 finished, for one reason or another: 'pregnancies, early marriages, just not attending classes, drug abuse or else', she thought. But then, after transition to the secondary education and becoming an adolescent, the issue of sexual relations and pregnancies rose again and had a slightly different tone in the research participants' stories.

There are commentaries on the type of educational institution, particularly at the post-primary level, arguing the pros and cons of single-sex or co-educational secondary education. As pointed out earlier, in general, the consensus in the literature is that girls in single-sex schools tend to perform better in national examinations than those in co-educational schools, particularly in science subjects and mathematics; there is, however, evidence of different experiences too, as was exemplified earlier. In this research, barring Tumaini, all other participants went to girls' schools for their secondary education, at least for their A-level studies; if the school where they did their O-level exams was co-educational, the hostels and boarding were for the girls only. Actually, as an interesting case, Rabia was enrolled in a single-sex school, a boys' school, because she was allowed to attend that particular school where her parents were teaching. She discovered that, 'for me, I think, it was the best school'. Tumaini was selected on the basis of her exceptionally good performance to attend a special school that was co-educational. She explicitly talked about the peer pressure of having boyfriends and sexual relations, and how she, together with a group of five other young women, was able to pass that period of adolescence without sexual relationships.

> Temptations were part and parcel of school life. However, I managed to finish my O-level without engaging myself into sexual relations. For me,

getting into sexual relation was associated with making parents sad and annoyed. I remember I had a boyfriend with whom we agreed not to have a sexual affair.

(Tumaini, 33)

In contrast to the other research participants, Tumaini talked a lot about the benefits of co-educational education and quite strongly discouraged schools for men or women only. For instance, Genefa suggested that the government should build more boarding schools for girls, which would help them to fully participate in education, perform better, and finally to join higher education. She and Naomi both referred to the safety issues regarding transportation in Dar es Salaam as an example. Genefa described how there were girls and women who credulously believed that some men were there aiming to help solve their problems (e.g. transport), 'but in the long run, finding them pregnating them'. Although exemplified slightly differently, Rabia raised a similar concern. She discussed the risks that the poor girls and young women coming from the villages faced. If boarding school or hostels for girls were not an option, they ended up renting a house, but they did not have enough funds to maintain all of the costs of housing. They did not have enough funds to take care all of the schooling costs either. Maybe the young women also thought, 'I have my free-dom for the first time' and then if there were men around, maybe they said, 'yes, what will you give me'. In other words, Rabia explained how the girls and women were dating and having sexual relationships in order to finance their schooling and everyday living costs; how they were not outright sexually harassed or forced into sexual relationships by anyone but the circumstances. 'Forced by the situation' is how Amana described the same phenomenon.

It's like when the girl is not getting enough from home, she may decide to be offered to any man, regardless of the status of the man, like the age or anything. [...] And that girl is getting chips or sweets and money sometimes, and get involved in the sexual relations.

(Amana, 36)

For secondary education, Amana was enrolled in a girls' boarding school. Although they were allowed to go for shopping one day per month, and they had one day for visitors, she suspected that some of the pregnancies in her school were actually caused by the male teachers. This extremely difficult and delicate issue of harassment and abuse of girls in schools is examined in studies led by Leach (Leach et al. 2003, Leach and Mitchell 2006), and touched on by Stambach (2000), and one of the participants in this research. Naomi's experiences regarding inappropriate behaviour of a male teacher will be looked at more closely in the following section, which focuses on teaching and learning experiences and encounters with teachers.

Along with Amana, 'the one free Saturday in a month' was referred to by Wema, which she described as being so busy with going to the shops and eating different food in comparison to the school, etc., that she did not give the slightest thought to dating and having a boyfriend. Still, the sexual behaviour of girls was under observation in her school. The girls used to have pregnancy tests three times a semester: 'even before I had my periods, the nurse would come and check your breasts and press your stomach', explained Wema. She, as did many of the research participants, admitted that she actually did not understand how you get pregnant, because they had never had that kind of conversation with their mothers: 'we never discussed at home about reaching puberty, menstruating, and boys; no ... my mother said "why you don't ask your teachers to tell you what to do"', Leyla recalled. Then she continued with how her mother would have 'swallowed her alive' if she had found out that Leyla had a boyfriend. Boyfriends were secrets (from parents and the school personnel) and 'things to do with boys and sex', were learned from friends and fellow students at their boarding schools: 'we misled each other', Leyla concluded (cf. Sommer 2010). Wema, as did most of the research participants who stayed in boarding schools, told stories 'of serious inspection' not only regarding sexual behaviour, but concerning private issues and intimacy:

> At 7 o'clock, the matron of the hall will pass by and check the cleanliness of your bed, how perfect you have made your bed (...) I remember she even used to check your underwear and stuff...
>
> (Wema, 32)

Leyla reflected that, on one hand, the threat of being suspended from the school for good scared the girls enough not to mess with boys, and on the other hand, the religious ambience at the missionary schools had the same result. Still, in her school, one girl died in the dormitory when she was trying to get an abortion, and another tried to get one, but she failed. Leyla knew she was allowed to return to school and sit the formal examination, after she had written a letter where she asked for forgiveness, because it was her last year in the school. Yet in the first place, she had been sent away because she was a bad example for the other girls.

As said above, not all of the harassment in and around the schools was caused by male pupils, nor was all of it sexual in nature. However, it is important to note that not all of the experiences and memories regarding 'others' were bad at all; on the contrary, most of the school memories told by the research participants were, indeed, pleasant, and related to their fellow pupils.

'The Place You Go and Play'

> My first memory from school, it's like, it was a place where you go and play. That's the memory I have. Yah, it was like this playing thing.
>
> (Wema, 32)

Wema's story is slightly exceptional, because she started schooling ahead of time. The reason was that her parents did not want her to stay at home alone,when her brother who was one year older started going to school (see e.g. Lewin 2007). Therefore, for Wema, schooling was associated with playing. Similarly, when I asked about nice school memories, basically all of the research participants talked about playing with other children. There was not that much leisure time for the research participants: even in Saturdays, the first thing to do was to wash the school uniforms. But then after, as mentioned by Tumaini: 'whatever little time you get', you could go and play, and at school, there were still the breaks. Amana, Genefa and Wema all talked about how they had enjoyed the breaks at the school and just being and playing with their fellow pupils; 'I was really happy about the friends', Wema concluded. Many of the women also talked about netball, athletics, dancing club, etc., in other words, being active and doing something, *functioning* with others, during the recess.

To continue with the positive memories, in contrast to the bad experiences regarding boys, some of the research participants gave examples of co-operating and playing with 'the other' sex. As explained by Rabia, because there were animals on the way to school, boys actually protected the girls who lived far from the school. According to Leyla's insights, nobody harassed them on their way to and from school, but instead, the teasing was more like playing, like the boys just running after the girls. Amana's reminiscences included occasions when they invited boys from their school to play with them, and 'to change experiences and ideas'. In addition, she held pleasant memories about travelling home for holidays by train or with a special bus booked by the school and reserved for the students only.

For Rehema, as for being disabled, the assistance and support from her friends was of great importance; similarly, Genefa and Tumaini talked about the encouragement from other female students. As we remember from Genefa's story, she received very little help from her parents with her education and schooling. She had to work a lot to support her older sister to take care of her children and the rest of the siblings. Because of all the domestic work, she did not do that well at school. Genefa had been one of the best students in the school where she started her primary education, but once she moved to Dar es Salaam to live with and help her sister, her results collapsed: 'I was very much disappointed', she recalled. But then, she joined Std VII, 'worked very hard with assistance from my friends', and was selected to join one of the most famous and prestigious girls' schools in the whole country. From Tumaini's story, we remember how important it was to her to have a group of 'like-minded friends' regarding 'good manners; without boyfriends or men', as she puts it. For Tumaini:

The best memory (in secondary education) was friends. I remember, I made lot of friends. We were having a group of five ladies who were doing well; very close to each other and we were teaching each other. [...] We were together during recess and after classes, playing, joking, and yah.

(Tumaini, 33)

As was expected on the basis of the policy-research informed pre-understanding, and evidenced in the women's practice–perspective narratives, factors that have an influence on girls' and women's participation in the school system and classrooms go beyond the school infrastructure and facilities (only), and are interlinked critically with human relations. Relating to 'others', boys and girls, was discussed previously; the research participants' experiences and insights concerning encounters with their teachers and teaching and learning practices follows.

Learning Environment

In addition to the physical school environment, the learning environment affects female performance and attainment in schools. As pointed out earlier, the literature reviewed by Odaga and Heneveld (1995) suggests that teachers' attitudes, behaviour and practices have perhaps the most significant implications for female persistence and academic achievement and attainment. They also emphasise how teachers' attitudes towards their students reflect not only the academic capacity of girls, but also the broader societal comprehension about the role of women in society (cf. Fuller and Clarke 1994). Odaga and Heneveld (ibid.) found evidence of gender-biased attitudes and behaviour and practices, on the one hand, and other cases with little gender discrimination in class on the part of female teachers or their male colleagues, on the other. In this section, I examine first the research participants' experiences and insights concerning their teachers' attitudes and behaviour towards them, and the teachers' professional qualifications and skills as manifested in classroom practices. Then I look at the women's views of themselves as learners. In doing so, I pay attention to their self-images, which are mirrored, for example, in their subject choices, in their grades, transitions, etc. In other words, I examine their overall school performance and try to capture *the idea* they had about themselves as students.

The Issue of Teachers

When I asked the research participants to tell about their overall experiences concerning learning environments and teachers' attitudes, the relationships with teachers seem to have a lot to do with everything else except teaching and learning *per se*. The issues of (physical) punishment and strict discipline came up in all of the narratives. In the previous section, I touched on Naomi's experiences regarding harassment by a male teacher. These experiences are illustrated in the following women's stories. 'It's like this kind of funny relationship', sneered Wema, when I asked her to recollect teachers, female or male, in primary, secondary or higher education.

> There are so many memories about teachers in primary school. Like one
> teacher used to ask me to sell candies. [...] It is so normal, like every

teacher will bring stuff to sell, and give students to sell during class. So, one teacher used to give me candies to sell, and I had to balance sometimes, if I was not careful enough or maybe some kids stole it, then I had to compensate with my own money, which I would ask my mother to give me.

(Wema, 32)

Wema continued with how in Std V and VI in primary school her worst memories were related to beatings, 'sticks and caning'. She had a mathematics teacher who used beating as a method of learning, and Wema recalled being beaten on her buttocks 'all the time'. She said that she was almost ready to give up schooling for good but her mother refused to allow her to do so, simply saying: 'No, you are not dropping school.' Hanifa too, had a mathematics teacher who used similar kinds of methods: 'If you fail, you get one stick, if you happen to get five or six wrongs, then you get five or six sticks.' She described how most of the students used to hide somewhere, just to not be beaten by the teacher. This particular mathematics teacher of Hanifa also had an extremely biased idea of girls as learners:

He thought that mathematics is for boys and of course he used to tell us that. He used to tell us: 'You, come', 'you, do!' 'You think you can pass mathematics?' Maybe from that point, we (girls) learned that mathematics is for boys. And if it happened that one or two girls were able to answer questions correctly, he was like: 'Ha, these are boys, like men, you see, these are men'. And if a large group of girls failed, of course, he used to enjoy. He used to enjoy from his heart, you could see.

(Hanifa, 33)

Hanifa ended her story commenting: 'He was so bad; he's one of the teachers I hated.' She kept on musing on whether the fear of punishment, combined with the idea of appropriate behaviour for girls in the classroom, that is, silence, had the result that they were neither courageous enough to face male teachers nor to ask any questions. 'Why? Because you are a girl', she concluded. I started this chapter with Leyla's story, where she described how she used to run to school to be there in time to avoid being punished:

The caning! We were caned, girls were caned on our palms, and you know, you got caned until your hands turned blue. [...] Even in the class, if you got something wrong, for example, you got caned. And we were so terrified (holding herself) because of the punishment, caning.

(Leyla, 53)

As with Wema, Leyla told how her parents gave no alternative to a small child for going to school: they just said, 'You have to go there'. Leyla also conveyed the long-lasting anxiety of her primary school experiences:

We felt that because everybody else was punished, that's how it should be. But when I look back now, I think it was too terrifying. I got so scared. Even now, sometimes I dream that I am late for school, so I wake up and find my heart beating. [...] So, that was it, we learned, we had to learn the hard way.

(Leyla, 53)

Consequently, 'no more caning!' is what Leyla highlighted in her secondary school experiences. As for Wema, she remembered that sticks were still used; however, the girls were no longer beaten on the buttocks but on their hands instead. From Wema's story in the earlier chapter, we learned how strict the discipline was in her secondary school regarding sexual behaviour, and also that many other women in the study referred to 'inspection' and 'reporting time' and talked about 'prefects', 'watchmen' and 'matrons' of their schools. Even so, and despite the punitive disciplinary actions, Wema pointed out what a big and positive role the boarding school actually had for her, precisely because of the very strict matron:

I don't know... She just like... teaches you how to take care of yourself as a girl. As I said, my mother and I, we don't discuss like personal stuff, maybe something, but not so much how to manage my life as a woman or a girl.

If Leyla's and Wema's parents did not give any alternatives to going to school and taking the teacher's behaviour as it came, Naomi's parents decided to withdraw her from the school and transfer her to another, because of her being harassed by a male teacher. Naomi told how the teacher started to behave offensively towards her, after her refusal to become involved with him:

I said, no, I am your student. [...] From there, he started following me in everything I do. For instance, I used to sweep the road going to the staff room, so that the road was very clear. I am very hard working so I used to wake up very early in the morning and clean the road, and everybody could see that the place is clean, but when that teacher came to the assembly, he said, 'you see, this road is very dirty, it's not swept for weeks' and then he calls me in front of the whole hall and he used to beat me fiercely. That was the habit. [...] I became very bitter because he used to beat me very very hard.

(Naomi, 32)

Naomi described how things got even worse when she was selected as the head girl of the school:

Oh my goodness, it was very bad, the teacher used to pin me whenever something is not being done, so it's my responsibility. [...] That was really really hard time and that's when I went to my mum and explained this.

(Naomi, 32)

At that time, in Naomi's school of, there was 'no headmistress, no responsi-bility, no firm leadership', as she phrased it: 'teachers were missing; the first semester we did nothing; socially and environmentally the school was not good: food was not good, the hygiene was not followed, you find that even dogs and cats were eating from the same basin that we took, water was also a problem', she listed. All in all, the learning environment in that school was 'not conducive enough' for one to 'calmly sit down' and 'study effectively', as defined by Naomi, and that is the reason why her parents removed her to another school.

The benefits of recruiting female teachers into schools (positive role models and parental confidence about their daughters' safety), particularly in rural areas, were argued in studies reviewed by Odaga and Heneveld (1995). Although there is some evidence of the positive impact of female teachers on female student achievement patterns, some analysts suggest that there is not so much difference in educational advancement between the students of female and male teachers (ibid., Arnot and Fennell 2008; Casely-Hayford 2008). Earlier, Tumaini, for example, described how she had a really good secondary school mathematics teacher, a male, whom she was actually com-paring to a history teacher of hers, a female, whom Tumaini thought to be 'so cold; not charming at all in the class'. She also explained how she could not understand anything she was teaching, and as a result, ended up hating the whole subject – contrary to mathematics, which was her favourite subject. Leyla too, remembered having a very strict female teacher who used to cane the students much more often than the teachers in other classes. She told how she used to sneak to the neighbouring class where there was a male teacher, who 'was so kind'.

> Of course, she (her female teacher) used to know us, so she came (the class next door) with her stick, hiding the stick behind her and said: 'I've lost one of my sheep, you, you belong to my class'. And I was trembling, I was so terrified.
>
> (Leyla, 53)

Although significant space has now been allotted to the women's 'funny' and not so nice experiences with their teachers, it needs to be stressed that, as with 'relating to others', a great deal of their stories regarding learning environments and learning experiences were positive, and highlighted good and committed teachers, both female and male. These are examined next in their narratives considering the professional competence, pedagogical knowledge and skills of the teachers.

Based on the studies reviewed by Odaga and Heneveld (1995), there are both grounds and evidence for the following kind of argumentation: because of poverty, governments are unable to pay teachers' salaries regularly, which results in teacher absenteeism and a lack of motivation, causing further negative impact on the quantity and quality of the time spent teaching. This, in turn, degrades student performance and attainment. If one is not comfortable accepting the causal explanation regarding poverty and its relationship to the teacher's instruction time and attitudes and to students' learning achievements, a similar kind of logic and reasoning on the premise of what teachers actually value in their lives is, indeed, one conceivable explanation for poor quality in teaching and learning, as contemplated by Tao (2013; see also Towse et al. 2002; Wedgwood 2005). On the other hand, the very same studies referred to just above (see also e.g. Vavrus 2009) found evidence of motivated, committed and pedagogically ambitious and interested teachers. Likewise, the participants in this research emphasised that, despite the poor physical school environments and poor teaching facilities – and despite the pedagogically not so well equipped teachers – there *are* good teachers, and some really good teachers. 'My teachers were all good, I cannot say very much excellent, but they tried very much to keep on what they were doing', Amana said. Rabia's experience was similar, a totally different situation from her village primary school, where they got a new teacher coming straight from the teacher training college and with no teaching experience whatsoever. Still, she tried her best to make the classes work and the children to feel comfortable – by cooking tea and giving them bananas – although no one actually learned anything:

> Almost the whole class didn't know how to read and write, all what we did was to memorise the whole book [...] You say: A; but you don't know what A means [...] Children were very happy but she didn't know how to teach, so, you were there because of the tea and bananas.
>
> (Rabia, 30)

Interestingly, Rabia herself was given a class of students from Std IV to Std VII who did not know how to read and write to teach in the evenings right after she reached Form IV, which is the last grade at the O-level stage of secondary education. This, in a way, exemplifies and illustrates the urgency and need to recruit teachers, no matter if unqualified and without pedagogical knowledge and skills or any teaching experience at all.

It is notable that quite a number of the women's memories are attached explicitly both to the subject and the teachers of mathematics. Hanifa, despite the teacher she told of earlier, used to like the subject and passed easily until Form IV; Amana, Amisa and Tumaini really liked it as well, and, in contrast to Wema and Hanifa, they also liked their mathematics teachers: 'He was good, he was young, energetic and active; he was very much encouraging', they characterised them. Leyla described her mathematics teacher thus: 'He was very good teacher. He was so nice'. Likewise, Genefa, although not performing that well in mathematics, liked the teacher, because he was very committed and very encouraging, giving constructive comments. Genefa also thought aloud that if she had only had enough time to focus on studying, she could have performed much better. She reflected upon the importance of teachers in the learning results, saying that without this particular mathematics teaches, she could have had an F, but with him contributing a lot, she managed to score a D. The influence of teachers, whether you like the subject or not, and whether you perform well or not, was also discussed by Tumaini.

If quite a number of the women referred explicitly to the mathematics, many of them also talked about their language subjects and language subject teachers: Kiswahili, but English in particular, were raised in many of our conversations. Leyla pondered how difficult it is to prioritise the knowledge one gains from the school, but decided then that the language skills are the most important, because: 'through that we learn everything else'. Leyla had a very nice language teacher in primary school and she did really well in that subject, as did the whole class. She deemed that this one language teacher had a big influence on her later becoming a language teacher herself. She also recalled how this teacher, 'an old man', used to urge the girls not to be left behind the boys, saying: 'You should do better; work, you must work!' Apart from primary school, Leyla had a nice language teacher, from England, in her secondary school. She explained how they used to have many talented and committed teachers from abroad at the missionary school. For Hanifa too, the woman teaching language was her favourite teacher. She was a diploma holder, and, similar to Leyla's teacher, very encouraging for girls: 'If you study hard you will be like me; even I was like you, I thought some of the subjects were difficult, but now, I'm here teaching you.'

Yet, in general, in primary schools, Hanifa thought the quality of language teaching was pretty poor. As an example, in her school, English was not a medium of instruction but learned as an examinable subject. To clarify, her teacher happened to be Indian by origin, and Hanifa explained how she did not understand anything, and thus did not like the subject. As we recall, Leyla had a similar kind of experience from her advanced secondary education. When Hanifa began her secondary education, her school was a private missionary school where one of the rules was to use English only – and she did not even know how to make a sentence in English, being already in Form I. But then, she started to learn, started to like it, took that as one of her major subjects at A-level, in her diploma courses at the university, and was an English-language teacher by profession, when we met.

English language was precisely the subject that worried Genefa during her primary education, making her uncomfortable to use English 'even up to now', as she said when we met. They just did not have teachers in the primary school to teach the subject. On the other hand, the English subject teacher was one of her favourite teachers once she finally got English classes at secondary school. English language is what Hanifa, Leyla, Naomi, Rehema and Wema listed as their favourite subject. Why? 'Because I was doing better in that subject'; 'I tended to like subjects which I got good grades', they observed. Amisa wraps up the circular relationship between the (favourite) subjects, teachers' attitudes and performance in the following quote from her secondary school experiences:

I liked teachers who taught History, Geography and Languages, because I liked the subjects. I worked very hard, read on my own, sometimes consulted friends who were like two years ahead of me. You know, teachers tend to love those who are doing well. I think all were good teachers.

(Amisa, 26)

The ambivalent sentiments about English language in Genefa's case reflects her experiences regarding the quality of the schools. As mentioned earlier, she was selected to one of the best schools for secondary education, and for her, the contrast between the primary and secondary school teachers' qualifications and motivation manifested quite clearly:

Contrary to the former (primary), teachers (at secondary school in Dar es Salaam) were very committed. They (primary school teachers) were very lazy, they didn't attend the school and the performance (...) only one or two students passed to join secondary school.

(Genefa, 40)

Quite often, when I asked the research participants to tell me about their teachers at different levels of education, the most representative adjectives used to describe their teachers were 'nice' and 'good'. When I asked the participants to clarify whether they meant that their teachers were 'nice' as a person or as a teacher, 'good' regarding pedagogical skills, etc., it was only after this question that the issues of qualification, subject knowledge, pedagogic capacities, etc., overall experiences concerning teaching and learning as such, were brought into the conversation. At this point, Rehema told a different story as to what she meant by saying that the teacher was very 'good':

In primary school, I met friends: some were very good to me, some were not; they were like: 'wow, what the hell is she doing here, she is disabled' [...] the most memorable is my best friend and my class teacher. She was very good to me – as a person (clarified, after I asked). She said 'don't worry about this, you just work hard' and then she came to the class and announced that every person who will take advantage of this girl, he or she will be punished. [...] Then she educated the other people to know that disabled people also have equal rights; that they should be with us, together, so as to be close; that we are all from God. Then my colleagues understood. [...] She was a very memorable teacher. I got support from her.

(Rehema, 28)

An interesting detail as to why Tumaini and Wema both felt that they were treated 'nicely' by their teachers (the exception being Wema's mathematics teacher) was that 'they were kind of known' because they were teachers' children. Wema considered that she had a good relationship with her teachers, especially the women, even some of the men, but still she did not *like* male teachers. According to her experiences, there were clear gender-biased attitudes on the part of male teachers that affected even the student's' grades. She gave an example of dissecting a rat as group work in a biology class: despite the fact that it was done together, the boys got higher grades than the girls did. Amana referred to a similar kind of experience and worried how much girls depend on what and how they are being taught in the classroom: if the teacher is not 'aware' of all the children; if girls are not involved; if they are not answering, etc. Obviously, all these kinds of experiences regarding teachers, female or male, who try or try not to 'keep on very much' in the 'profession', if they are not all trying to 'excel', impact on learners' self-images, their expectations, motivation, and their overall school attainment. These aspects of learning are investigated in the following section.

Women's Self-Images as Learners

As expounded upon in earlier chapters, the research participants had performed really well in their education. Earlier, it was also elaborated how these ten women had benefitted from performance- and examination-based advancement. Their self-images as learners, which are to some extent reflected in their attitudes towards subjects (choices), as well as their overall educational performance connoting grades, completion, transitions; droppingout/grade repetitions/ failures (if any) etc., are at the centre of this section. Nevertheless, as may be surmised from its title, the overview drawn from the women's stories is highly affirmative:

> In primary school, I was very lazy but I was very intelligent. I didn't care but when I reached and passed Std VII, I changed. I learned everything and I was really happy.
>
> (Rabia, 30)

> In my class, I was among the best. [...] I enjoyed science and mathematics, because I used to answer several questions right.
>
> (Amana, 36)

> I remember in English, I used to beat all the boys. Then you feel like, yah, you did it! The first person is somebody!
>
> (Wema, 32)

Wema pored over her perceptions and wondered if the boys actually felt the same: 'I am a boy and I got higher grades', or if it goes without saying that as a boy you should do better than girls. She, however, felt strongly that part of her success came, not only from the good grades *per se*, but from being the best, and better than any of the boys. Likewise, 'being the best' was what Rehema figured to be her best memory from the school: she was one of the students doing well in the class and 'that was fun!' In addition, she was the best student in essay writing in the whole school and was selected to represent the school in a competition. Genefa similarly enjoyed being in her class in primary education, and notably in stream A, because 'most students thought that stream A has bright students'. She was bright herself, and in fact, her nicest school memory were associated with the results and transitions, completing Std VII in primary school, in particular, with excellent grades, thus enabling her to enter the famous secondary school in Dar es Salaam, which was against her sister's prediction of her not passing, because she thought Genefa played too much. Identical to Genefa, Rabia's nicest school memories are intertwined with excellent results in completing Std VII and being among only three students from her school to enter secondary education. As we learned from Amana's quote above, she too was among the best. Indeed, she and another girl in her primary education were the ones who performed the best, then five boys followed. Tumaini was ranked to be number one or number two in the school, and Hanifa also performed well, being the best in her class, or if not that, then the second or third.

Hanifa told, crestfallen, how the memories of being an excellent student and doing really well at school, were at the same time the best and the worst, which reminded her of the sister who did not do so well. This story is closely connected to Hanifa's 'familial enabling environment', which will be presented in detail in the next chapter. The above suggests that all of the women in this study were doing exceptionally well, starting from the very beginning of their educational careers – apart from the 'laziness' of Rabia in her early school years, and the fact that she attended classes 'because of tea and bananas':

> In Std II, I didn't know how to read and write [...] I memorised the whole Swahili book in my head [...] when I reached Std III, a new teacher came and told my mother (that she couldn't read and write). Then, in the evening, I was beaten and I learned to read and write in two or three days...
>
> (Rabia, 30)

Although what was uttered above gives a very positive picture of the women's experiences regarding overall educational advancement, were there any failures or setbacks at all in their educational careers, as might be implied from Hanifa's remarks? Generally speaking, in primary education, all of the women performed well and transferred from one grade to another smoothly, without repeating or dropping out. Genefa, as an example, managed to advance without severe hiccups, regardless of the challenges at home. Even so, she was not very happy with her grades, precisely because of the challenges at home: 'My grades were not bad but I couldn't score high', she said. She explained by giving an example from mathematics learning: 'Maybe I could have managed mathematics, but I did not have enough time to do the exercises [...] and, you know, mathematics you can do better if you do a lot of exercises, if you practise.' As we remember from Genefa's earlier narratives, she was obliged to spend a lot of time queuing to get food, fetching water, taking care of her sister's children, etc. Therefore, clearly, in her case, there was not enough time to invest in schooling. Wema was performing well in every subject except mathematics, but she was ready to drop out of school for good because of the mathematics teacher, as discussed earlier:

> It was really really horrible. You just lie on the floor and hold your skirt together, then the teacher will cane you as much as possible, and then you just go home swollen and crying. I don't know, but mother will just tell you like, 'Do your exercise' or 'Improve!' Still, I didn't get good grade in mathematics.
>
> (Wema, 32)

Wema deemed that without the determination of her mother, she would have quit. She also reckoned that she did not meet any of her primary school mates during her educational career, and she did not know whether any of them had dropped out along the way, or if some, or any, had continued. She did remember one boy from next door who had repeated year after year after year: when Wema finished her primary education, this boy was in Std V and they had started together; what happened to him afterwards, she did not know.

However, Wema also repeated one class. To be more precise, she was made to repeat a class, because she had started schooling one year in advance, as we remember from her story earlier. When she and her brother, who was one year older, reached Std III, it was decided that they could not be in the same class and therefore Wema had to repeat. 'I remember I was really unhappy', she said. Like Wema, Hanifa also was made to repeat one class, and she too, for reasons other than poor performance. At Std VII, which is the last year of the primary education, her sister got sick and was admitted to hospital. Her sister used to tell Hanifa and everyone, especially the doctors, that she, Hanifa, being there would help her to get cured. Hanifa stayed with her for about six months and naturally was left behind in her studies. As a consequence, she repeated one class. Leyla told of an interesting episode when her father decided that she needed to repeat a class. She said that she made such a fuss about it that her father reconsidered and decided to let her continue. 'Maybe he thought that because I am short and tiny I cannot finish, and I'm not able to continue; maybe he thought that I should learn more, so that I could do well in exams and beat the class', she recollected.[5] However, Leyla did really well, even without a repeat. Rabia failed once, in Form IV in secondary education, but her failure, as with that of Wema and Hanifa, was not due to her performance. In her case, there was an instance of exam cheating in the school and consequently, all of the students were penalised.

Amana got married right after the Form IV examination and then had a baby. When the results came, she found out that although she had performed really well in the subjects earlier, she had not passed mathematics and chemistry. As a consequence, Amana was not able to continue, but stayed at home for four years instead. After that, she divorced her husband and managed to go back to school. The story of Amana will be looked at in more detail later. Similarly, because of family reasons, Hanifa had to postpone her university studies. The role and meaning of her husband and family in constructing her educational career will be examined more in the following chapters. In addition to familial reasons, Hanifa had health problems that conditioned and caused ups and downs in her university studies. At university, Wema, 'to her surprise', did not get a good degree. She almost got 'third lowest class', which is just marked 'pass' and she did not find a proper reason or explanation for not doing well there. In comparison to everywhere else that she had studied: 'maybe I was socialising too much, maybe I relaxed somehow, my mother wasn't there pushing me any more, there were no timetables, too much freedom for me at that time... I don't know, I attended classes all the time but I didn't work extra on my own', she recalled.

Generally, from a gender perspective, Genefa, Rabia and Leyla considered that in primary school, the boys were doing better. Genefa said that it was predominantly boys, 'of course, were performing very nice as compared to girls'. This was also Rabia's experience: 'women were failing while trying to impress the whole society that they are ready for marriage [...] They were told that if you don't work hard (at home) you won't get married, so, no purpose for schooling [...] if you know how to read and write, what more do you want from school?' Leyla's opinion was that in primary school, some of the girls dropped out just because they had had enough: 'I was too young to know why, but when I look back now, I think they preferred to get married [...] They went back and joined the village life, sort of.'

According to Amisa, boys were also doing better in secondary education and she gave three reasons for that: the first is the way the boys and girls are raised and treated according to 'Tanzanian culture', the second relates to poverty in developing countries, and the third concerns the relationships between females and males. She was quite upset as she explicated these issues by saying:

> You know, I was able to work hard because I was not like an inferior object. I was treated equally with boys since my childhood. But if someone has been raised like 'you are a woman, you belong to kitchen' you will always behave like you belong to kitchen. [...] Also, in the Third World countries, all the problems seem to affect women more than men. [...] And according to my experience, those girls who were having relationships with boys were performing very poor.
>
> (Amisa, 26)

Amisa gave two examples regarding relationships between girls and boys, women and men. The first was more like a 'childish love story' while the second concerned a more serious kind of (sexual) behaviour. She described how some girls just laid in the dormitory beds during prep time and naïvely daydreamed about the boys who had just told them 'I love you, I miss you' stories; at the same time, the same boys were actually at class or in their dormitory working hard and reading. Also, Leyla talked about the issue and impact of having a boyfriend at secondary school: 'no sex at all but just having boyfriends; we used to write letters to each other; but then my boyfriend left me for another girl and I was so discouraged'. The second example from Amisa included some misconduct and was, according to her, 'a really really problem'. She asserted, 'Let me finish':

> Our school was at the city centre, you just walk like ten minutes and there were those businessmen. And because of financial problems, they (her schoolmates) had their boyfriends at the city centre and at the same time their school boyfriend. So, you can understand how difficult it is to handle two relationships. You have to go for that guy at the city centre for money. [...] there were others who were having relationships outside, so, during the weekend is the time when you sneak out. Or maybe, your man will come during night and then you sneak out.
>
> (Amisa, 26)

'You can understand what I am talking about?' she wanted to confirm. Amisa, as did Wema, had a really persuasive and restrictive mother, and because of her, Amisa worked really hard to be an outstanding student throughout her educational career. In fact, Amisa conceded that she was afraid of her mother, to some extent; she, being really competent, and very strict, would sometimes lock Amisa in her room, pinching her and shouting, 'why don't you understand', in particular in trying to teach science subjects to Amisa at home.

The science subjects and mathematics were again raised by many of the women with regards to their grades and performance. In other words, they did not perform well in those particular subjects: although they did not fail in exams, repeat classes or drop out of school because of the subjects, the subjects were referred to quite passionately by most of them. Rabia depicted her father as having been kind of known and respected in the mathematics, chemistry and biology subjects, but no one in her family actually excelled and chose the subjects as their combination, and for Rabia in particular, being the oldest child in the family, 'it was a shame', she confessed. In secondary school, it was only Tumaini who liked it very much and used to do well in science subjects. In Form III, there were around ten females in her PCM (Physics, Chemistry and Mathematics) class and about 25 males. She was supposed to continue with the same combination because of her excellent results in mathematics, but then her uncle decided that it was a subject for males only, and instead, Tumaini should take biology for her major. This affected her learning results, because she was not good in the combination of physics, chemistry and biology (PCB), and later also her university choices and career:

I am not good at cramming, and for biology, you need a lot of cramming. I was good in mathematics, more than in biology, but since I agreed, I had to work hard to continue my PCB. [...] Then came the second year and I didn't perform well in exams, I got division two. I started to apply for the university, but since I didn't do very well in biology, I missed the opportunity to join a medical college [...] Then again, the uncle advised me to take home economics and human nutrition. For him, it was a course that was probably suited better for women. I agreed with my uncle.

(Tumaini, 33).

Apart from certain points of their educational careers, discussed above, when Leyla (after being transferred to another school for her A levels) and Naomi (because of the school environment and the teacher) did not do well in their education, they and the rest of the research participants were all happy with their grades – and for good reason, because they all got excellent marks. For the most part, they all described being pleased with their overall school performance, the learning outcomes and the knowledge and skills that they gained. Furthermore, regardless of school and learning environments that were not always so enabling, they had gained a rather good *idea* of themselves as learners, and the positive self-image was even emphasised after the acknowledgement of 'I know it's my own work', depicted by Amana. Still, in order to be 'among the best', 'being intelligent' alone is not enough; still, as emphasised by Tumaini 'good points don't come just like that, you have to work on them'. She, as an example, considered secondary education as being the hardest time in her life, requiring a lot of effort to succeed; extra efforts were paid in particular at the time of exams, when she recalled sleeping only for a few hours. Indeed, the uppermost *idea* that I gained from the women's narratives of their school experiences and insights is that they were truly working hard, all the time, as is concluded in the following section.

'Yah, We Used to Work All the Time'

In this chapter, the research participants shared their experiences and insights regarding school attendance and encounters with fellow students and teachers. Thus, in addition to getting acquainted with the research participants' physical school environments, the learning environments and human relations have also been examined in detail. The *idea of Tanzanian woman* that arises from the research participants' narratives is somewhat encapsulated in the statement by Wema: 'Yah, we used to work all the time.' On the other hand, she summed up the school experiences that most of the women shared as she sighed: 'I think all my childhood was beating.' In these two quotes, Wema raised two themes that were discussed by all of the women regarding their 'participation in' school.

The research on education, gender and development is packed with literature discussing the importance and interlinkages between education, gender and *poverty*. Even though they did not define their families as being poor, all of the participants in this research also emphasised financial constraints on pursuing education; conversely, they saw their standard of living to be moderate and 'okay'. From the women's schooling stories above, we learned that for their primary education, they were enrolled in government schools, and for secondary education, they had been either selected to join government secondary schools or admitted to private ones. Though the government schools were free and other costs relatively low, all of the women's financial situations had been tight, and they were assisted by their (extended) families. In addition, all of their families and all of the women had been involved in some way or another in generating extra income to finance their education. Evidently, educating children, girls and boys, had been demanding and an extensive investment for the research participants' parents and other family members. Yet the direct and indirect costs of schooling did not prove to be prohibitive for these women.

From the women's stories, we heard that they were top-performing students. That partially explains why they were not too heavy a 'burden' for their parents when it came to investment priorities and decision-making. Thus, as learners, the research participants were really proud of their achievements, and for a reason, because neither the physical nor the learning environments in their *primary* schools were enabling enough to actually open up a set of capabilities to function. Instead, on the basis of their narratives, we may conclude that the quality of both the physical and the mental environments were rather poor. For most of the women, the schools were quite a distance away, the water and toilet facilities were there but the quality was poor, the food was non-existent or bad, the classrooms were poorly equipped and fights over desks and chairs were normal. Both female and male teachers frequently used corporal punishment as a method of learning as well as to maintain discipline. Furthermore, the teachers held somewhat biased ideas towards girls' ability to study and learn. Consequently, for most of the women, the difference between their primary and *secondary* schools was striking, regarding both the facilities and the teaching and learning ethos and atmosphere. Although the boarding secondary schools differed significantly from the research participants' primary school experiences, particularly in the sense that 'for the first time' they were well and truly able to study, still they were not only just sitting and studying in a 'conducive learning environment' but instead, as Wema discovered, they were really working hard all the time.

In addition to the work-loaded school days, 'the beating' seemed to characterise many of the women's school experiences and insights into their childhood. In fact, the issue of corporal punishment was raised powerfully by most of the women, and that is why many rather bad images were attached to their teachers. On the other hand, the research participants also had some very good teachers who significantly influenced them to construct their educational pathways. Similarly, despite fights in the classrooms and some difficult incident with the boys, the women held remarkably warm memories of their fellow students. Finally, many of the deficient experiences and the low quality of education were comprehended 'as normal' and 'as things should be'; hence, enabling and supportive *enough* environments to stay enrolled and pursue further education.

The above practice perspective, voiced by the women, to complement the policy-research informed knowledge base and understanding of the complexities of girls' and women's education and schooling in the Tanzanian context, is the essence of this research. For that reason, significant space has been given to the women's story-telling to draw pictures of their school environments, on the one hand, and to capture their experienced and perceived school-related practices, on the other. Women's attendance in primary and secondary education was examined closely for two reasons. First, the early years and experiences from school seemed to be pivotal, not only for further academic advancement, but also for future career prospects, and more importantly, for the self-image of the women. Second, these are the levels of education on which the recent policy perspective and initiatives in Tanzania are focused, and accordingly, to which the policy focus of my study is directed. Yet, to reach an understanding of the women's school experiences and insights not only as educational pathways, facts *par excellence*, but to indicate their well-being and agency, entails a capabilities-informed analytical application being used next.

Well-being and Agency in Schools

It would be somewhat easy to draw an overwhelmingly long list of factors that constrained the research participants from functioning and pursuing goals that they have reason to value on the basis of their narratives. But instead, and in accordance with the purpose of the research, the focus of this last section concerning enabling school environments is on factors that *supported* women in constructing their educational well-being and agency. I am not framing the whole data set discussed above with the capabilities approach but instead am extracting the evident critical issues that arose from the women's narratives.

The key assumption in my study has been that the women had a reason to value education and hence to pursue to achieve that goal. As a matter of fact, the presupposition was that the women had valued and ranked education highly and invested a lot in education (social, human and economic capital), as being representatives of a tiny proportion of Tanzanian women who have reached the university level of education. Thus, to start with functionings, as one aspect of educational well-being: what have the women achieved? They accessed education in the first place, and attended school on a regular basis; they were able to learn, perform well, transit from one grade to another almost without interruptions (excluding a few repeats); they completed compulsory education (primary) and further basic education (O and A levels in secondary education); thus, they were able to 'fully' participate in formal education. To continue with functionings, the women continued with their studies, some of them directly to university, most, however, at first to the teaching profession (Teacher Training College); afterwards they were enrolled at university, first to pursue a Bachelor's degree and later, at the time we met, for their Master's (barring Rabia). Consequently, their well-being achievements contrast with the *idea* of girls' and women's education and schooling in the developing countries, including Tanzania, characterised by the functionings of dropouts, grade repetitions, poor performance, etc., as discussed earlier. Undoubtedly, the women's functionings exhibit a rather easy and smooth educational pathway if assessed as a retrospective compilation of well-being achievements; if, however, the focus of analysis is directed to their capabilities, which are their opportunities to realise the various functionings, we end up evaluating their well-being freedoms, which were rather less easy, as with where their school environments were concerned. Real opportunities and the genuine presence of serious options that are concrete and available to the agent are importantly embedded in the notion of capabilities. Similarly, the availability of concrete basic amenities such as water, food, desks, chairs and books, to name but a few, may have an influence on how the students convert the available resources into functionings (learning) at school. However, there is no causality between the availability and amount of resources and equality, nor are they equivalent to people's well-being, that is, what the learner has achieved (functionings) or could achieve (capabilities). Instead, the capability means an opportunity that is made feasible (enhanced and/or constrained) by both internal (personal) and external (environmental and social) conversion factors. Hence, what kinds of sets of capabilities were open to them, enabling them to function?

As discussed previously, within the primary and secondary education reforms in Tanzania, the improvement of school infrastructure and learning environment is given a high priority. However, the availability of desks alone may not advance individual well-being (enable learning), if, for example, and as we heard from Leyla and Hanifa, a female student is to be seated between two boys to keep discipline and calm the classes. Hanifa, together with her sister, was able to come to an agreement with their teacher that it was okay for them to sit together. Leyla, in comparison, did not: maybe it was not in accordance with the school rules and classroom practices in Leyla's case to ask for any concession; or maybe she did not have the personal courage to face the teacher and speak out; as we heard, she was honestly afraid of some of her teachers. Therefore, the availability of the very same resource may or may not advance the achievement of educational functioning (e.g. learning), depending on social (e.g. school rules) and/or personal (e.g. courageous) conversion factors. Let us consider another example, concerning the availability of books. As we learned from the women's narratives concerning primary schools, very few of them had textbooks of their own. The books at the school were for school use only. Furthermore, in most cases, the books were shared, among as many as nine students. Even the exercise books were stored at the school in some cases, because the teachers did not give any homework assignments, knowing (or presuming to know) the familial conditions, which, in most cases, did not enable the students to study at home. Hence, the availability of learning resources, even limited, did not advance the functioning of learning, because the potential of learning materials was not fully utilised for learning purposes. Thus, in principle, the set of capabilities within the school environments of the women included important basic resources such as desks and books to be converted into functionings. However, there were also social conversion factors, such as the *ideas* of the teacher concerning the motivation and abilities of the students to study, or the seating arrangements according to gender, embedding the idea of expected and appropriate classroom behaviour *inter alia*, which, in practice, limited the research participants' opportunities to function. This is not to claim that the resources and facilities at the schools are of no relevance; quite the contrary: as pointed out by Tumaini, for example, to have the opportunity to use the library, and have access to books and newspapers and other learning material correlated indisputably with her well-being and achievements. However, to realise the utilisation of opportunities in Tumaini's case was dependent on her mother, who worked at the same school as a teacher. Hence, resources are only a means to advance the achievement of well-being – presuming that they are adequately utilised, not indicators of well-being or development, when assessed through the lens of the capabilities approach.

Nussbaum (e.g. 2000, 78–80) envisaged a list of (ten) central human capabilities, including basics such as bodily health (inclusive of being adequately nourished), bodily integrity (being able to move freely from one place to another; having one's bodily boundaries treated as sovereign) and affiliation (referring to social bases of self-respect and non-humiliation *inter alia*; this resonating within the capabilities approach in general with the idea of 'the ability to go without shame', as per Adam Smith). These three essentials are evidenced and mirrored in the women's school-related experiences. First, the issue of water and food, that is, 'adequate nourishment', which are evident and concrete amenities to enhance children's opportunities to function: to concentrate better and indeed, to learn. Some of the research participants mentioned that they might have had some tea at home in the morning and/or maybe tea or porridge at the school, but most of these women went to school and also returned home 'with empty stomach'. The lunch (if) provided at the schools was found to be poor, and in Leyla's case, supposed to be paid for by the parents, which occasionally turned out to be prohibitive for schooling as well. Most of the women recalled not having any water during the school day, or if was available, the purity and hygiene seemed to be questionable. Thus, in accordance with Nussbaum's definition, these women's early school experiences are suggestive of missing one of the very central opportunities to achieving educational well-being.

Second, to have access to school, literally, that there are schools 'close by', which for the participants in this research meant 'five kilometres', 'taking two dala-dalas', or just walking 'next door', is clearly one of the critical prerequisites to achieving educational well-being. These women did not identify any such environmental and/or social factors that would have constrained them from concretely reaching their schools – if 'mountains' and 'harsh conductors' are not taken into account. Still, the women narrated some disturbing behaviour and encounters with the male students, and in that sense, they did not have the opportunity to 'move freely from place to place' or have 'sovereignty over their bodies'. Furthermore, as we learned from the women's narratives, one reason for their early awakenings on school days was due to the long distance, whether living in rural villages, towns or in the city of Dar es Salaam. However, the second reason to awake around 5 o'clock in the morning was in consequence of their fear of being late to school and punished. Most of the women gave examples of how afraid they were of their teachers, hiding from them and wishing not to be paid any attention:

> You know, in primary education, this is a joke, I don't know if you do it, if a teacher asks a question and you think, I'm not ready to answer this question, you do this (crossing fingers) and imagine that your teacher might forget you.
>
> (Hanifa, 33)

Corporal punishment seemed to be used in the women's primary schools as both a disciplinary action and a somewhat common teaching method. Obviously, the 'stick and cane' methods are in flagrant contrast with the ideas of bodily integrity, self-respect and non-humiliation, and they also neglect the duty of 'child-friendly pedagogy', which is today one of the key components of education sector reform in the country in general and in teacher training in particular.

To comprehend schools as 'social bases of non-humiliation', and to secure 'bodily boundaries' interlinks also with the already discussed availability of resources such as desks and chairs. Most of the women described how they shared desks; some fought for the chairs and if not 'successful', they sat on the floor. Wema recalled how upset she was at ending up on the floor and soiling her nice clean skirt; likewise, Leyla remembered how the dirty or torn uniform was a reason to be punished (in front of the school) and excluded, thus constraining her opportunities to participate in school. Resources are clearly an important means for the students' educational well-being. However, even more meaningful is = examining what they can actually do with them: for example, most commonly it was the girls, because they were not expected to and/or 'allowed' to fight for the chairs (social factors), who ended up sitting on the floor, and so by the end of the day, they were not wearing an impeccable uniform, which is not acceptable according to the school rules (social factors). Furthermore, as discussed earlier, the parents and the girls themselves tend to be rather sensitive for modesty reasons about their appearance (social factors). Thus, despite the evident availability of resources, various intersecting factors may constrain girls' and women's opportunities to convert them into functioning. Finally, referring to Nussbaum's central human capabilities, there is no *de facto* reason or justification to value schools and classrooms as any different kind of environment than the social bases of self-respect and non-humiliation securing sovereign bodily boundaries.

I wish to give one more example of resources and the importance of personal abilities in the conversion of resources into well-being achievements. In the following excerpt, Naomi talks about her primary school experiences as a small child in a boarding school and the usage of such a small and seemingly insignificant issue as the bucket.

M-A: So please tell me about your first school memory?

N: We used to work, oh, the work was very heavy, you can imagine […] functions of my bucket, and everybody had their buckets. We were living in double-decker beds like this, up and down, so everyone's bucket is here or there (pointing to the edges of the bed in our room at Onnela). With the bucket, one would carry water with to the kitchen; with the bucket we use it to wash yourself and wash your clothes; with the bucket you have to take manure from the shamba (field) maybe to the gardens; with the bucket you have to take the flour from the milling machine to the kitchen; with the bucket sometimes we have to use it as a toilet in the night, same bucket; with the bucket sometimes we have to take sand for brick making or mud, because if you have to dig, and then you water it and stamp on it several times, so that you get mud for brick making.

So you can imagine, with the same bucket we used to do all these things! Because it was a plastic bucket, if a bucket breaks, you have to buy another one immediately. That's why we've had pocket money at the sister.

So you can imagine all that work is done by girl or a child of 10, 11 or 12 years and also we used to carry timber from the saw machine, yah. So we were working a lot.

I am telling you, when we were at school, we even used to make bricks, yah, we made bricks. We could do everything: we knew how to keep cattle; we knew how to grow vegetables; we could even slaughter a pig. Can you imagine, girls at that level?

M-A: (…) you have been telling me about your memories, the hard-working times, what do you think, are they good or bad memories, are they the best or the worst?

N: Well, I think it's not bad […] By then, I thought it was bad because it hurt me a lot, but now I think it is good, because it is from that base, from that foundation that I became a brilliant woman. (eehh) Right now, I can do everything in my own house, because if you know work, it is easy; you won't hesitate or be afraid to go anywhere, because you can manage the environment.

Evidently, apart from the academic knowledge and skills, Naomi's school had a particular kind of an idea concerning the competence that the girls ought to embrace, either to assist in maintaining the school and/or for the benefit of the girls themselves, presumably future mothers and wives responsible for running their own households. Regardless of the reasoning, the key point here is that there might be significant variations between persons in the conversion of resources into functionings. As we learned from Naomi, she needed to have the bucket in order to function in the school in the first place. Thus, to achieve the literally various beings and doings, having necessarily nothing to do with educational functionings, the set of capabilities had to comprise the bucket. But, in addition to ownership of that particular asset, she needed to be endowed with physical and mental personal characteristics to make use of the resource according to requirements. Let us consider that instead of Naomi, it had been Leyla or Rabia who was attending that particular school: necessarily Leyla, being tiny in physical size, as she defined herself, or Rabia, being physically impaired, would not have had the stamina to utilise the resource and function. Besides, the opportunities to realise functionings in Naomi's case definitely required a particular kind of mindset, enabling her and her fellow students to cope with such a hard-working and demanding school environment. As is heard in Naomi's discourse, she asked me many times : 'can you imagine'; 'so, you can imagine', and I used to reply 'no, I really cannot imagine', because to me, due to my own experiences and insights, schools for children aged between 10 and 12 are truly more like 'the place you go and play' and Naomi's and her fellow students' personal resilience just keeps amazing me.

Apparently, it is easier to list and discuss the factors that constrained the women from constructing their educational well-being than to identify supportive factors. However, the potentially prohibitive issues, such as the distance to school, the lack of (poor) food, non-existent water, the poor classroom conditions and learning facilities were comprehended in the women's narratives as 'normal', 'usual', and so 'we coped', as Rehema phrased it. In that sense, regarding physical school environments, the barriers were not such big problems as to have prohibited the research participants' schooling or to have caused dropouts, grade repetitions, poor performance, etc. (characteristic of girls and women's functionings in developing countries). Instead, the school environments might be regarded by the parents as embracing 'sufficient conditions' to actually keep their children enrolled at school, as noted by Sayed and his colleagues (2007) in a study on educational exclusion and inclusion in South Africa and India. As a matter of fact, regarding primary and secondary education, it was factually the parents of the women who had reason to value education, whatever that reason and reasoning being, whether intrinsic ('prestige', family tradition), instrumental (later financial support for family) or both. Consequently, the opportunity of *not* being enrolled and participating in school in the first place was not available to them. In that sense, the research participants' parents were acting properly according to the claims made by Sen and Nussbaum that, even though the agency freedoms of the child might be diminished, they should be required to remain in compulsory education because of their incapability to make mature value judgements and 'imaging' the future benefits of education.

The majority of the women got a different perspective on 'normal' and 'usual' upon entering secondary education and the kinds of school environments that had appeared unimaginable earlier. Comprehended in this way, all of the prohibitive and constraining factors, previously discussed, may be reversed, and indeed, regarded as supportive factors in the context of secondary education – in accordance with the primary intention of the focus of this analysis. Rehema's remark above is suggestive of at least some kind of an adaptation and 'inability to imagine' difference, and for that reason, the expressions that implied discontentment of the women regarding facilities and resources in their primary schools came from the statements that actually concerned their secondary education. 'It was so different!', meaning 'no more caning', 'all girls', 'good teachers' and 'friendly environment', as noted by Leyla, as she described the private missionary boarding school in which she was enrolled. This implies how she and many of the other women perceived and judged the school and learning environments of their primary education, simultaneously identifying a set of opportunities to enable the realisation of their well-being achievements in the secondary education.

It also implies what the research participants truly had a reason to value. As was narrated by the women, they were excellent students and if they performed worse than anticipated, they were really disappointed. Hence, it was in accordance with their *agency goals* to learn and perform *well*. We also learned how they appreciated and enjoyed having the opportunity to 'calmly sit down' and 'study hard and effectively', possibly for the first time in their lives. Thus, their agency goals were prerequisite agency freedoms to study and learn, and to perform well in a *conducive and harassment-free environment*. Hitherto, it was expounded that as per the capabilities approach, the well-being of every person is the ultimate goal of human life. Correspondingly, Robeyns (2005, 2006) suggested that the assessment of well-being should focus on the question of whether a person is being put in the conditions in which she can pursue her *ultimate ends*, hence referring to substantial agency freedoms. Retrospectively enlarged upon, we may conclude on the grounds of the women's explicit definitions: 'I was (amongst) the best', 'I was the first', etc., that was able to reach the particular level of attainment that they had a reason to value. But instead of the obvious agency achievements, how did they manage to realise the valuable beings and doings: what kind of factors actually comprised the conducive and harassment-free environment supporting them to act as per their aspirations? According to Leyla, it was good teachers, no more caning, and all girls.

It was not just Leyla though: most of the women gave several examples of remarkably good teachers who made a big impact on their lives – some during their primary education, but most often after reaching the level of secondary education. Amisa, for example, was convinced that the teachers were 'trying their best. They are committed.' All in all, the women's experiences and insights are suggestive that, for the most part, they had good, encouraging, talented and qualified teachers in secondary education, enabling them to perform well, and pushing them to work hard and do even better. In addition to the teachers' significant support for the women's academic attainment, the research participants narrated how the influence of some of the teachers was so empowering for their comprehension of being a female. This was both explicitly, as with some of the teachers and matrons of Leyla, Wema and Naomi; and implicitly, as in Naomi's school giving all the duties and tasks that the students were assigned on top of their academic work. Some of the practices that the schools maintained might be comprehended to be against women's agency freedoms, such as 'working hard as it was hurting me a lot', as told by Naomi or the inspections regarding personal hygiene and pregnancies, which are not all in line with the idea of bodily integrity, for example. Yet the women did not convey these practices to be violating their agency freedoms as such, since they agreed them to be reasonable, warranted and justified because of what they saw happening around them. Furthermore, on the grounds of some of the women's remarks, it may be claimed that the strict discipline in their schools actually had a positive impact on their agency achievements.

From the women's primary school experiences, we learned how some of the teachers had extremely gender-biased ideas of girls as mathematics learners, for instance, and for which reason they were quite systematically silenced in the classrooms. These kinds of practices and social arrangements are against the freedoms of the girl, presuming that she values the capability of expressing herself. In the women's secondary schools, slightly more opportunities to exercise this agency freedom was given; indeed, one might claim that it was the time of schooling when the first seeds of giving voice for oneself were sowed. In addition to the *social conversion factors* of 'school ethos' and teachers, there were friends and fellow students who critically enabled the women to exercise their agency freedoms and supported them in pursuing their goals. Wema gave voice to this in stating how she 'really enjoyed' and how 'happy' she was just to *be* with and *do* things with her friends. Genefa valued her friends helping her to study hard and catch up from being left behind due to the quality of her primary school and demanding family conditions. Rehema too, gave credit to friends for assisting her, for example, to move around in the school compounds, but even more importantly, for their attitudinal support. Tumaini talked a lot about the group of like-minded friends; and Leyla, for her part, appreciated the girly conversations, no matter that they were actually 'miscounseling' each other. Obviously, apart from the fellow students support to academic agency achievements, the women's friends played a great role in the process of identity formation.

Apart from the overall supportive ethos of the women's secondary schools, personal characteristics, critical for the women's educational well-being and agency are evident. First, one can argue on the basis of their achievements that they were intelligent and talented students. Despite the limited resources and opportunities provided by their first school experiences, they were able to perform well and 'to be selected' into good secondary schools and one of the very prestigious ones, where their personal abilities and academic ambition were supported and enhanced. However, they also worked hard, valuing the idea of being the best and the first – being not at all pleased, for example, with the achievement of 'being enrolled' only. Obviously, given the limited substantive opportunities to function, particularly in primary education, it is clear that a certain kind of mindset is needed: first, to set agency goals, and second, to act accordingly. The difference between the research participants' primary and secondary school experiences and insights is axiomatic and rather striking. This is not only regarding the availability and amount of resources and opportunities to function, but also concerning the actual achievements in relation to the agency achievements and the freedoms they had in pursuing their goals. Given the various constraining factors reflected in the women's school-related sets of capabilities, it is presumable that the research participants' social and familial environments comprised opportunities that enabled them to function and enhanced their agency freedoms, to be examined next in detail.

Notes

1 200 TSH (Tanzanian shillings) = 0.1€.
2 For comparison purposes: a taxi ride within the city area would cost 20,000 TSH, and a loaf of bread 800–2,000 TSH.
3 'Providing a meal at school is a simple but concrete way to give poor children a chance to learn and thrive' is emphasised within the NEPAD initiative piloted in nine African countries, not only helping the children to concentrate better and, indeed, learn, but also benefitting their households, which are at least a bit relieved of having to scratch for food.
4 Interestingly, mathematics, in particular, was referred to by many of the research participants, including Leyla in her story. I will discuss subject-related matters in the later section. See also FAWE/FEMSA (Female Education in Mathematics and Science in Africa) http://www.unesco.org/education/educprog/ste/projects/girls%20africa/femsa/femsa.html
5 See Lewin (2007), who discusses parental decision-making regarding being enrolled into a certain grade despite the 'official' age of the children.

Bibliography

Arnot, M., and S. Fennell. 2008. Gendered Education and National Development: Critical Perspectives and New Research. *Compare: A Journal of Comparative and International Education* 38, no. 5: 515–523.

Brock, C., and N.K. Cammish. 1997. Factors Affecting Female Participation in Seven Developing Countries. Education Papers No. 9. 2nd ed. London, UK: DFID.

Carr-Hill, R., and J. Ndalichako. 2005. *Education Sector Situation Analysis: Final Draft Report Revised.* Dar es Salaam, Tanzania: Economic and Social Research Foundation.

Casely-Hayford, L. 2008. Gendered Experiences of Teaching in Poor Rural Areas of Ghana. In *Gender Education and Equality in a Global Context: Conceptual*

Frameworks and Policy Perspectives, ed. S. Fennell and M. Arnot, 146–162. London, UK: Routledge.

Colclough, C., S. Al-Samarrai, P. Rose, and M. Tembon. 2003. *Achieving Schooling for All in Africa: Costs, Commitment and Gender*. Ashbourne: Ashgate.

Colclough, C., P. Rose, and M. Tembon. 2000. Gender Inequalities in Primary Schooling. The Roles of Poverty and Adverse Cultural Practice. *International Journal of Educational Development* 20: 5–27.

Connell, R. 2007. *Southern Theory: Social Science and the Global Dynamics of Knowledge*. Cambridge, UK: Polity Press.

FAWE. 2001. *In Search of an Ideal School for Girls*. Nairobi, Kenya: FAWE.

FAWE. 2003. *The ABC of Gender Responsive Education Policies-: Guidelines for Developing Education for All Actions Plans*. Nairobi, Kenya: FAWE.

Fuller, B., and P. Clarke. 1994. Raising School Effects While Ignoring Culture? Local Conditions and the Influence of Classroom Tools, Rules, and Pedagogy. *Review of Educational Research* 64, no. 1: 119–157.

HakiElimu. 2005. *Three Years of PEDP Implementation: Key Findings from Government Reviews*. Dar es Salaam, Tanzania: HakiElimu.

King, K., R. Palmer, and R. Hayman. 2005. Bridging Research and Policy on Education, Training and Their Enabling Environments. Special issue, *Journal of International Development* 17, no. 6: 803–817.

Leach, F., and C. Mitchell, eds. 2006. *Combating Gender Violence in and around Schools*. Stoke-on-Trent, UK: Trentham.

Leach, F., V. Fiscian, E. Kadzamira, E. Lemani, and P. Machakanja. 2003. *An Investigative Study of the Abuse of Girls in African Schools. Education Research Report No. 54*. London, UK: DFID.

Lewin, K.M. 2007. *Improving Access, Equity and Transitions in Education: Creating a Research Agenda*. No. 1 of *CREATE Pathways to Access Series*. Brighton, UK: CREATE.

Madamombe, I. 2007. Food Keeps African Children in School. NEPAD Supports School Feeding Programmes. *UN Africa Renewal Online*, October 4 2012. http://www.unorg/africarenewal/magazine/january-2007/food-keeps-african-children-school.

Nussbaum, M. 2000. *Women and Human Development: The Capabilities Approach*. Cambridge, UK: Cambridge University Press.

Odaga, A. and W. Heneveld. 1995. Girls and School in Sub-Saharan Africa, From Analysis to Action. World Bank Technical Paper No. 298. Washington D.C.: World Bank.

Posti-Ahokas, H., and M.-A. Okkolin. 2015. Enabling and Constraining Family: Young Women Building Their Educational Paths in Tanzania. *International Journal of Community, Work and Family*. http://dx.doi.org/10.1080/13668803.2015.1047737.

Reinharz, S. 1992. *Feminist Methods in Social Research*. Oxford: Oxford University Press.

Robeyns, I. 2005. The Capability Approach: A Theoretical Survey. *Journal of Human Development* 6, no. 1: 93–114.

Robeyns, I. 2006. The Capability Approach in Practice. *Journal of Political Philosophy* 14, no. 3: 351–376.

Sen, A. 1992. *Inequality Re-examined*. Oxford: Oxford University Press.

Sommer, M. 2010. Where the Education System and Women's Bodies Collide: The Social and Health Impact of Girls' Experiences of Menstruation and Schooling in Tanzania. *Journal of Adolescence* 33: 521–529.

Stambach, A. 2000. *Lessons from Mount Kilimanjaro: Schooling, Community, and Gender in East Africa.* New York and London: Routledge.

Tao, Sharon. 2013. Why Are Teachers Absent? Utilising the Capability Approach and Critical Realism to Explain Teacher Performance in Tanzania. *International Journal of Educational Development* 33: 2–14.

Towse, P., D. Kent, F. Osaki, and N. Kirua. 2002. Non-graduate Teacher Recruitment and Retention: Some Factors Affecting Teacher Effectiveness in Tanzania. *Teaching and Teacher Education* 18, no. 1: 637–652.

UNICEF. 2012. *The State of the World's Children 2012: Children in an Urban World.* New York: UNICEF.

Unterhalter, E. 2003b. The Capabilities Approach and Gendered Education: an Examination of South African Complexities. *Theory and Research in Education* 1, no. 1: 7 22.

URT. 2001. *Primary Education Development Plan 2002–2006.* Dar es Salaam, Tanzania: Ministry of Education and Culture.

URT. 2003. *Joint Review of the Primary Education Development Plan (PEDP): Final Report.* Dar es Salaam, Tanzania: The Ministry of Education and Culture.

URT. 2004. *Joint Review of the Primary Education Development Plan (PEDP): Final Report.* Dar es Salaam, Tanzania: The Ministry of Education and Culture.

URT. 2005. *Poverty and Human Development Report 2005.* Dar es Salaam, Tanzania: Government of Tanzania.

Vavrus, F. 2009. The Cultural Politics of Constructivist Pedagogy: Teacher Education Reform in the United Republic of Tanzania. *International Journal of Educational Development* 14, no. 1: 65–73.

Wedgwood, R. 2005. Post-basic Education and Poverty in Tanzania. Working Paper Series No.1. Edinburgh: Centre of African Studies.

9 Enabling Social and Familial Environment

Despite the numerous potentially prohibitive factors at the school level and the social and familial levels of the enabling environment, the participants in my study were all enrolled in school and they were not withdrawn; they did not drop out, and their academic performance and grade level attainment was good, exceptionally good for some, as was evidenced in the previous chapter, which focused on the women's school environments. Here, the analytical interest is in factors related to the social and familial levels of the enabling environment, elaborating the research participants' familial arrangements and everyday life practices, on the one hand, and the views and attitudes of their parents regarding the relevance and value of education and schooling of their children, on the other. I am not postulating that the sets of capabilities in the women's familial environments did not entail problems and barriers to the education and schooling of the children, particularly girls. However, given the level of attainment and education that the women have achieved, I am presuming that there were more supportive practices and ideas towards education and schooling of children in their families than in many other families around them. However, before embarking on a closer examination of the research participants' narratives on a 'typical day' and their familial attitudes towards female education, I wish to give voice to Amisa and Hanifa to tell their stories to depict some of their familial experiences concerning schooling. The reason for starting by listening to their stories is that a great deal of the women's significant school memories addressed human relations and relationships with others. Indeed, the most intense memories of school were intertwined with their teachers and peers. At home, understandably as children, the most intense memories were associated with their parents: with Amisa, her first and worst, and for Hanifa, her best and worst school memories were associated with their fathers. A couple of times, Amisa mentioned how painful it was for her to go back to such bad memories, but even so, she told her story:

> My first and worst school memory is not very good. I don't want to remember, but I have to share it with you. That time we were still with our father, and my father, in my opinion, was very bad man. I don't want to even remember... My father was very harsh, very rude, who didn't

want to see you. I mean like 'traditional African man' who are like, when they are coming, you have to run and hide, because Dad is around.

I was joining Std I. Then you are just telling him, you know, 'Dad, I'm going to school, are you going to buy me a uniform?' You are like a kid and he didn't speak anything to me. I remember that and I was six years old. My mother was cooking and she said 'yes, yes, you are going to school'. Then my father came and sits and said: 'I've told you, I don't want to see you (the mother) to go to work, but you said no (she wanted to work); and now you tell your daughter to come and tell me to buy her a school uniform. I'm not going to buy anything!'

And because my mother knows this guy, he's going to beat her, so let's just keep quiet, but you know what he did? He just took a big bowl, glass bowl, and threw it to my mum. It cut my mum here (shows her head) and then suddenly my mum just dropped and fainted. I can't forget that because my mum was bleeding [...] I had blood in my clothes, it was terrible (Amisa is crying). I shouted and then neighbours came and picked my mum and took her to the hospital. Till today, my mum has an injury in her head.

Beyond doubt, Amisa's father was seriously and violently misbehaving. Conversely, Hanifa's father did not have bad intentions; rather, he was inconsiderate and unintentionally inflicted severe consequences on the whole family and on Hanifa and her sister in particular. For Hanifa, the best primary school memories were related 'for being the best – or the second or at least the third', but at the same time, they reminded her about her twin sister, who did not do that well at school:

My father told us that whoever will become the first one in class will get present and anyone who will not become on top ten, she or he will get punishment. I remember that day, when we were given our report at school, I knew I was the first one; therefore I reached home to show my dad the report. But in doing that, I didn't know that I was affecting my sister who was not in top ten. From that day, the problem arose. My sister is suffering some sort of ... like ... what do you call it ... some sort of depression and it started that day! And I, my father and my mother we know that! You know, I was young. I was not thinking about my sister, I was thinking only about me. Therefore, I rushed home quickly to show my father that I was the first one; my sister didn't run home, but she was hiding somewhere. [...] Even now, I regret it and my father regrets it, too. So, from that day my twin sister was not able to share things, maybe she feels she's not as important as I am, and from that day up to now, she's sick, she's got mental problems.

With the narratives of Amisa and Hanifa, I wanted to illustrate the power and emotionality of memories that the women held in relation to education and schooling from the very early years of their educational pathways. To keep this in mind, next, however, the research participants' personal experiences are analysed from a more pragmatic standpoint, by which I mean Berger and Luckmann's kind of 'everyday life practices'. Hence, in the following sections, I examine what the women actually did before and after their school days, presuming, on the grounds of their stories regarding school environments, that during the school days they were attending classes on a more or less regular basis. In the previous chapter, I highlighted how long and work-loaded the weekdays were for the women. In this chapter, I look in detail at what actually belonged in their days, apart from schooling as such. Were they able to 'calmly sit down and study effectively in a conducive (familial) envir-onment' or were they 'working all the time'; and if they were working 'all the time', was it only girls and women in their families who did that, or to what extent and how did the boys and men participate and do their share? In other words, in the following sections I am looking for an answer to a question: were there socially constructed gender roles and ideas in their familial set of opportunities that were either enabling or constraining them to function?

'Us'

In line with the epistemic positioning taken in the study, I am arguing that the factors of poverty, rural/urban settlement, household-family composition, to name but a few, are socio-economic and socio-cultural 'facts' and resources that impact on people's functionings. However, their gendered impacts, if any, are socially constructed, and influence the ways the girls and women operate, or to be more precise, are *capable* and *enabled* to operate. For example, the time allocation in households has a direct bearing on girls' and women's educational well-being: if only they are given the responsibility for carrying out a vast number of the household chores and other familial duties, at the cost of school assignments, for example. As Leyla accompanied us earlier into her primary school, let her now lead us from the school back home and begin depicting and further elaborating on her 'typical day':

> At 4.00 p.m., we were allowed to go home after singing a farewell song. I normally played on the way home with my friends. And often I would get home a bit late. My mother would scold me or cane me if I didn't give sufficient reason for coming late from school. Normally, there would be something for me to eat before I rushed with a bucket to the stream to fetch some water. I would go there two or three times depending on the need of water in the house. Then I would help my mother prepare the supper or wash the dishes as the case may be. It would normally be dark by then. After supper (around 8.30 p.m.) we listened to some stories from my mother on a day when she was not too tired or too angry with my father for spending too much money on local beer. Then we would fall asleep, only to be awakened by my mother at 5 a.m. the next day.

On the basis of Leyla's experiences, I start de-constructing the research participants' typical days and enabling familial environments from the viewpoint of poverty and standard of living. This is not to prioritise the issue of the socio-economic position of the families, or to explain and reduce the social and familial practices to poverty; it is merely to contextualise and put into perspective the intersecting issues of social class and gender roles and identities (see e.g. Fennell and Arnot 2009; Morley, Leach and Lugg 2009; Unterhalter 2012).

'We Were Not Very Poor Compared to Other People'

As previously discussed, educating girls entails both direct and opportunity costs of schooling, which are prohibitive to families, in particular poor families and rural families (Odaga and Heneveld 1995; Vavrus 2002; cf. Colclough, Rose and Tembon 2000). All of the research participants considered their families 'middle-class' and they 'were not very poor compared to other people', as Rabia said. Yet financial constraints were emphasised by all of them, even those from relatively privileged backgrounds, and different solutions were sought to fund schooling (see e.g. Posti-Ahokas and Okkolin 2015). After her parents divorced, Amisa was raised by her mother, without any kind of support from her father. Her mother worked as an accountant, but to supplement her salary, they had to sell cattle ('pre-inherited' from her parents). Apart from financial contributions, Amisa and her two sisters and brothers were helped out in many ways by their grandparents (on the mother's side). Wema's story is surprisingly similar to Amisa's, regarding the divorce of her parents, no support from her father, but help from her mother's parents and other relatives. Their stories are congruent also in the sense that their mothers were really committed and put all their efforts into ensuring that all of their children were enrolled in schools, attended classes and completed their education. Both of these mothers invested a lot of money in tuition, for example, to make sure that the children were learning and performing well, in order for them to continue and study further. Both Amisa and Wema got good grades and were selected to attend government schools, which their mothers could afford, and for university studies as well. Wema, for example, was granted a scholarship;but there were also books, exercise books, uniforms, socks and shoes to buy, and pocket money was needed for personal utensils like soap and snacks. Although Amisa and Wema both performed well and were selected to pursue further studies, given stipends, etc., the grades of Wema's siblings were not that good. Consequently, their mother had to pay everything: 'yes, certainly, definitely, it was hard financially to educate children', said Wema. Whereas Amisa's mother sold cattle to earn some extra money, Wema's mother used to bake chapatti (bread) and sell it in the staff room and cafeteria at the school where she was teaching.

Apart from the story that Amisa told above, she did not talk about her father at all. Wema referred to hers occasionally, when explaining the details of his non-supportive behaviour. As an example, he had suggested several times that Wema could come and live with him: 'If your mother doesn't have money (to buy warm clothes, as Wema was asking him to buy a new coat), you can come and live with me'; 'Me, when I was studying, I was struggling; even you should struggle', he had answered and refused to give any money for the children, despite his good job and salary. Wema thought that many of the problems in their parents' marriage were actually caused by financial matters: she explained how her mother had also taken care of the family alone earlier, when her father was studying for his Master's degree in the United States. At that time, however, there were only the children and the mother living together. When her father returned, a number of his relatives moved to Wema's home and they expected her mother to take care of everything, not only to cook food for the whole extended family after returning home from work, but also buying the food: 'I'm using all the money to feed this clan of people!' she had complained. Wema recalled an occasion when there was almost a fight over the food with their uncles, and the children were telling their mother that they were still hungry. She got really mad and shouted: 'There are so much people in here; I just want to cook for my children!' Wema commented that it was out of the question that Wema's father and his brothers would have cooked; instead, they were all waiting until evening for her mother to come home from work, and then go to the market and shops to buy the food and then prepare it. Another remark, just to give an example of their standard of living and what 'middle class' meant in Wema's case, is that it was only very recently, when she was at university, that they had their first television; until then, she had amused herself by singing, dancing and playing – much more than reading, as she noted. All in all, Wema recalled how she saw her mother really trying to make the available money be enough: 'she said that there is only this (amount of money) and you could really see that there is only this.'

Amisa's mother also struggled to make the money suffice. Amisa joked that her mother was very clever and gave only the minimum amount of money for her at boarding school to cope for one month. Amisa described how she and her sisters and brothers used to write a list a week before going to school and put everything they needed, even cookies, in the list. Then, in the evening when their mother came home from work, she said, 'okay, bring your papers', and then she started to remove what she considered to be unnecessary, 'this yes, this no, this no...' Then during the weekend, they went to the shops and bought necessities according to their mother's choice. 'You wrote and she chose', Amisa ended her wish-list story. Still, she thought her mother was fair because she distributed equally to all of the children depending on how much she happened to have at that time: 'Okay, you take this and then after a month, I'll come and visit you.'

In most of the research participants' families, there were two breadwinners, both mother and father, and in some families more, since the older siblings were also contributing. As with the parents of Amisa and Wema, the parents of Rabia and Rehema were also divorced. Their fathers were remarried, and Rabia and Rehema had both lived with their fathers' new families. Rabia's father worked as a teacher in a secondary school. He held a university degree in biology and chemistry; his new wife was also a teacher, and helped Rabia in particular with her language studies. Interestingly, two of her four sisters worked as teachers, and her grandfather and biological mother were also teachers.

Rehema was living with her father and step-mother when we met. Both of her biological parents were teachers by profession, but her father also made a living from his business. Rehema's stepmother had worked in a teacher training college, but at the time of our conversation she was a housewife and had a small business in handicrafts and clothing. I am talking about *two* breadwinners, knowing that, in general, it refers only to the main supporter in the family. The reason for doing this is that, although in many of the research partici-pants' families, the mothers were not employed outside the home, but were housewives, they worked in the family fields and gardens, kept the cattle and/ or did some petty trading. Thus, they worked and earned for the family needs, and the income that they brought to the family cash flow was crucial for many of the families, for example, for Hanifa's.[1]

Her father worked as a district education officer, and because of his occu-pation, their living and housing conditions were relatively good according to Hanifa's assessment. They had a big family: Hanifa had four sisters and two brothers, and in addition, their grandmother and uncle lived with them. It was her father who decided and took care of the financial matters in their family, and the funds from the small business that her mother had went to small things to run the household. However, the extra that came from her mother actually enabled the family to invest a bigger share of her father's salary in education, and for that reason, Hanifa judged, they were able to be enrolled in very good private schools. Indeed, because of the number of children, and the determination of her father to select the best affordable schools, the lion's share of the family income went to the schooling costs of the children.

In Naomi's family too, all of the four daughters were enrolled into 'very good' Catholic missionary boarding schools, as she explained, and their parents 'were really really struggling hard' to expend a lot on education. Similarly, all the sisters were 'struggling and striving hard to study', because they were taught that 'without education life is harder; education is the key!' Both of Naomi's parents were educationalists and both of them were retired at the time of our conversation: her mother from the profession of executive educational officer, and father after 35 years of being a priest; Naomi's mother held a BA and her father a PhD degree. One of her sisters also had a doctorate, and two others were studying at the university. Naomi perceived that they had lived in 'educational surroundings', because of the educational and professional backgrounds of their parents, but also because her parents maintained an education centre that provided non-formal education for people who, for one reason or another, were not able to follow formal educational pathways (compare Posti-Ahokas and Okkolin 2015). Naomi said that their parents worked extremely hard to support their family and educate their daughters; even today, 'they are scratching themselves', she continued, because her sisters did not have any scholarships for their university studies. Earlier, 'in good old days', as Naomi put it, apart from her father's dayjob, he used to work as a consultant. In addition, Naomi's family had a shamba, where they grew maize, bananas, oranges and mangoes for sale. They also had chickens and kept cattle, cows and goats, and got extra income from selling eggs and milk. Altogether, Naomi described the standard of living in their family as satisfactory, medium: 'We had our basic facilities, we could get our food, shelter is there, and clothes; we could see our doctor.'

Whereas Amisa and Wema were significantly supported by their mothers, and Hanifa and Naomi by their parents, of the participants in this research, Genefa, as noted already, was the least supported educationally by her parents. Her mother was a housewife without any formal education and only some adult education; thus, she barely knew how to read and write. Although her father had previously worked as a police officer, he had been in an accident and was forced to quit paid work. Genefa described the severe marks the accident had left on her father, and the whole family for that matter, because afterwards he did not take any responsibility on familial issues whatsoever: 'he was just too drunk ... he used to drink very much'. Consequently, Genefa, along with her sister and two brothers, went to live with their oldest sister. Although their mother helped, for example, by looking after the children of Genefa's sister, who by then was a single mother and working, Genefa also had to do her share to help support the family and personally take care of her own schooling and education. For example, after school Genefa went out selling bread and bananas. During the weekends, she went to the wholesale market in the morning, and then sold her small purchases with a minor profit in the evening. Interestingly, Genefa considered the petty trading to be one of the nicest and most fun things that she used to do with her sisters and brothers. When I asked about 'things to do with a family', she mentioned, 'just sitting together, maybe singing along the songs from the radio', as the first, and the small business for the second.

As for Tumaini's parents, they educated three extended family members alongside their own four daughters, no matter that neither of them was well paid, as she observed. Her mother was a primary school teacher and her father a prison officer – 'a job which is for people who have not gone to school, like our father', as Tumaini phrased it. Because of her father's work, they used to live in the prison camp staff quarters. To add to the family income, Tumaini's mother was very enterprising: sometimes she used to keep chickens and sometimes she had cows; sometimes she did small business activities such as roasting nuts and selling them at the school where she was teaching; there were also times when she bought fruits and sold them. She recalled that the 'not normal' behaviour of their mother to try to raise extra money was quite embarrassing for them, and that they were teased about that, but 'we coped with the situation', she concluded. Tumaini described how life had had ups and downs in a relatively poor family, but all the basic things were taken care of. She also thought that, maybe, because there were only four (biological) children in their family, that was the reason their parents managed to send all of them to school; maybe, 'taking care of us (only) was a bit easier compared to other families, where a lot of money has to go for food and other things, like sending their children to school'. In addition, she remembered that they were all doing well, and were selected to government schools, which the parents could afford. But still, life was not that simple, she summarised. Thus, despite her parents' low educational background, they had taught Tumaini and her sisters that: 'what will make you a better person tomorrow is education. You say, I am suffering today, but I will be relaxing and happy tomorrow.' Tumaini felt that financial constraints along with 'cultural issues' hindered many children from going further in their education but her parents placed a high value on education.

Whereas Genefa was very much involved in labour to promote the economic resources of their family, Wema's mother did not want the children to get into her small business. Thus, their mother used to wake up very early in the morning, bake the chapattis and sell them by herself at the college; or whenever their neighbours asked Wema or her siblings to sell them some bread, they called 'Mum, they want bread.' Wema explained that in their primary and secondary schools, it was quite common for students to come with bread or peanuts or bananas to be sold during the breaks, to add to their family's income, but they never did that in their family. However, they participated in many domestic duties, as did all of the other women in this study. The numerous household chores on which the girls and women in Tanzania, as well as across the developing world, spend a lot of their time, were listed earlier. In the following section, these duties are examined in relation to the research participants' everyday life arrangements.

'I Spent Most of My Time in Doing Domestic Activities'

> I don't know, we felt somehow that we had to help our mother. I knew how to make my bed and other stuff […] when I was still in primary school, I used to make my own food: when we came from school, you still have to cook the evening meal for me and my sister and for mother; there was not so much cleaning, but cooking, just normal stuff.
>
> (Wema, 32)

As discussed earlier, apart from the direct costs of schooling, high opportunity costs prohibit the education and schooling of girls significantly. Child labour for agricultural, domestic and marketing tasks is indispensable to some households, and girls are more likely to be involved in childcare than boys. In rural areas, children spend more time working than those in the urban areas, although the demand for domestic labour in urban areas has also increased. As a response, the rural households have sent their daughters into the domestic labour market to work as 'house girls' (servants) for kin and/or non-kin families (Odaga and Heneveld 1995, 17–19). Amana, to give an example of the research participants' experiences, was sent to help her aunt in Dar es Salaam, finally, after 'several refusals' by her father: 'She (aunt) kept insisting [...] she had two children, the third one was born when I was there. I used to play with them, take them to school and back, before I went to school.' Occasionally, Genefa's sister had a house girl to take care of her three children and at some point their mother was looking after them. Yet, according to Genefa's memories, when she was in secondary school, she spent most of her time doing domestic activities, 'every day', 'as usual':

> As usual, after school I went back home and helped my sister to do domestic duties. It was my duty, every day, to prepare dinner and also fetch some water. I also helped to do cleaning, etc.
>
> (Genefa, 40)

Thus, although Amana served in the 'domestic labour market', before she went to Dar es Salaam to assist her aunt, she had been fully occupied with agricultural and domestic tasks; she, as a big sister and the second oldest in the family, was allotted almost all the domestic duties that her mother and other women were doing. So, after school, if their mother was not around but working in their shamba, Amana did what her mother had assigned her to do: 'I had to make sure that all my siblings had eaten, I had to cook; we had to fetch grasses for our cows, and we had to do cleaning', she explained. The time to which she is referring concerns the primary schooling. Amana described how she 'as a child, I enjoyed reading but I had no time after school'; when at secondary school, she came home only for holidays: 'but the problem again was time, I had too much work to do', she recalled. Rabia too, was talking about the holidays, which were, in fact, not that much of a rest because their labour was needed in the family fields. Their mother was quite smart; Rabia laughed, and depicted how they were roped in for cultivation by the promise of a feast after the hard work:

> Then you work very hard and she goes to check: 'no, you didn't finish the other part' (laughs with Amisa), and then you work very hard the next day. 'But no, this part is still remaining', then you cultivate the whole shamba and maybe then she'll slaughter the goat. Then you just celebrate that day! And then the next day she starts saying 'we can slaughter another, if you finish this part (laughing) ...
>
> (Rabia, 30)

The lack of time, and consequently the lack of leisure, is the picture that is drawn from all of the women's stories. 'I rarely studied during night due to the fact that I had to wake up early to do some tasks at home, so, I had to sleep early', Genefa noted. We might also revisit what we learned in the previous chapter, with regard to the early awakenings, domestic duties and 'empty stomach', and get one more example from Tumaini who described her typical morning as including 'helping with house chores, like sweeping the outside environment, taking a shower, dressing up'. Tumaini did not remember if she used to have breakfast or not: 'I don't think so, or if there was any, then it was just black tea.' Hanifa shook her head, when I asked about 'nice things' that they used to do as a family: 'only, maybe, we are at home, maybe doing some duties, any duties, which we were given by our mother', she tried to recall.

Another thing that emerges from the women's stories is that the duties they were assigned were not considered only as hard work; on the contrary, quite often the agricultural and domestic work that they took care of with their sisters and brothers, or as a family, was comprehended to be one of the nicest things that they did together. Amisa recounted that the nicest 'family thing' for her was the time when their mother was building a house for them. She remembered that even the youngest brother was participating, and they were so dedicated to finishing their building project. Again, she recalled how the cattle that their mother had was a kind of family project; that first you make sure that the cattle are fed and comfortable, then afterwards, then you can sit at the table and eat as well. Furthermore, they used to go to their rice fields, which had been given to them by their grandparents and were mostly farmed by someone else, to help them and work there together as a family.

The third subject area that needs to be examined is the issue of gender roles in the research participants' families and how they varied significantly amongst them. As Leyla told her story, there was a clear division of tasks and duties assigned to either girls or boys. In contrast, Amisa gave an example of a family where gendered behaviour was not tolerated at all. Leyla explained how the main task for them as children, girls and boys, was to fetch water from the stream. However, boys only carried out this task up to a certain age. In other words, only young boys brought the water home. In their family, the girls, 'of course', as she said, helped their mother in the kitchen: they were fetching the firewood, lighting the fire, cooking and washing dishes; they were also cutting grass for the cows, because that was women's duty. Interestingly, it was the boys who were to feed the goats: because land was scarce, they needed to climb the trees, cut the branches and tie them together. Even today, Leyla mentioned that the division of labour in feeding cows and goats goes according to gender. From home, Leyla listed that she had learned lot of practical skills, such as cooking, washing and gardening, but first and foremost to 'be in time', second, 'not to be lazy', and third, 'to work hard'; if you were late, unnecessarily, you would be punished; if a woman is lazy, then the children will die of hunger, she had learned; and, if you are lazy and are not working hard, then you will be caned. She thought that the school system maintained and even emphasised what she learned at home. In contrast, Amisa narrated a story regarding domestic duties in which all the family members were responsible for and expected to conduct them, as appropriate to one's age and not gender:

Everyone was supposed to do something you could fit in according to your age. There was no like 'girls are supposed to wash dishes and cook', no, everyone were like 'today you sweep the floor' and 'you go and wash the dishes'. I remember, when we had long holidays, my mother used to send maids away, so, there was a timetable like my brother was supposed to cook every Wednesday; he was supposed to be in the kitchen... Sometimes he was saying 'Mum, I am a man' and she said 'What! You mean you are a man; have you seen any money in this house? I am the only man in this house...' so, it was like that. But, it helped us a lot. My younger brother is able to cook, wash the dishes, sweep the floor, I mean, we do all the work. My mum was very strict, very hard, so, it was like that.

(Amisa, 26)

On the basis of the experiences of Leyla and Amisa, we may draw two kinds of family portraits regarding gender roles: one that is in line with the rather conventional idea of the Tanzanian woman; and the other, which is quite radical in the Tanzanian context. Most of the experiences and insights described in the research participants' narratives may be grouped with the former, and only the story told by Wema is similar to the latter. To illustrate this distinction and to give exaggerated examples accordingly, Rabia depicted how in their village, according to their tradition 'men are supposed to bring home meat or fish every day, and the women are supposed to go home and prepare it; until, women were just sitting and waiting for the head of the house to bring home food'. We recall from Wema's story how her father and the extended family were waiting for her mother to come from work and start preparing the supper, after going to the market to buy something to cook. However, later, after the divorce, when Wema and her brother and sister were living with their mother, there were no gendered expectations regarding domestic duties:

For example, my brother, when we were living together, he was supposed to clean his own room [...] and cooking, yes, he was cooking. Sometimes he would come and wouldn't complain that 'why there is no food?' No, he would just go there (kitchen), get the pan and then call 'is there anybody who wants to eat?' And then we just go and eat his cooking

she recalled.

Wema was reasoning that her brother didn't want to maintain the kind of manly 'ego', because they had a single mother household. In the same way, Amisa gave an explanation for the unconventional household practices that they had, based on their mother's experiences regarding marriage and divorce, but also on how she had been raised. Amisa illustrated too how in particular to see her brother working in the kitchen was comprehended to be something really peculiar by their neighbours: 'they were like "why, why"...?' In addition, Tumaini mentioned her mother as being seen as rather 'eccentric' in the prison camp, where they lived because of her father's work. Tumaini recalled that where her father was concentrating only on his work, her mother was doing anything and everything to bring some extra income for the family and improve their standard of living:

> My mum was the one who was doing business to add to the income [...] The work which he was doing was kind of enough for him. And maybe, because other men in the camp were not really doing any other business to add to their salary [...] She was seen as not a normal women by everybody in the camp... sometimes they used to say, 'oh, she's Chagga' (ethnic group, famous for doing business).[2]

(Tumaini, 33)

Tumaini did not have any brothers, and evidently none of her own experiences regarding the familial and parental attitudes towards a gendered division of labour. Hence, she could not make comparisons to the boys as such, but like Amisa and Wema, she talked about the kitchen and cooking. According to Tumaini, it was consistent with their culture that the men were not supposed to be in the kitchen. Consequently, her father had never prepared any meal for her, and actually, she had never seen him in the kitchen. Yet, because they did not have any boys at home, during the holidays from the secondary school (boarding), there were many kinds of household chores for the girls to do, and 'no one would understand if you said, 'I am reading', as Tumaini phrased it. Naomi did not have any brothers, but in contrast to Tumaini, in her home, 'I'm reading' was very much encouraged, if not even insisted upon. Naomi described how her mother is her role model in working hard and coping with everything regarding domestic work. In contrast, her father is the example of how to read and study seriously: 'if you want to go to the study room and you get down to work, yah, we have a study at home [...] my father, what he normally does is sitting down and study ... even if you are in shamba, he comes and instructs "it's better you do this way..." but he doesn't do it himself, no.' Naomi laughed how they once nearly had a family crisis because her mother was travelling and he didn't have a clue how to feed the family: he couldn't cook, he could not even buy a loaf of bread; tea was okay and coffee, Naomi described how the only thing her father could do in the kitchen was to boil the water.

As previously mentioned, in most of the research participants' families they had an explicit idea of appropriate behaviour according to gender, including the idea of the division of labour. Wema told an anecdote about how, although her brother was supposed to clean his own room, he performed so poorly that Wema and even her mother volunteered to do it. Hanifa also mentioned how there were some attempts in their home to get the boys to clean their own places but they just simply failed to do that: whether unintentionally or purposefully, knowing that it was not really a boys' task, she could not say. For the most part, the boys and men in the research participants' childhood families did not attend to household work. Hanifa and Amisa described their familial arrangements: the girls and women held the main responsibility for that: 'we have learned that from our childhood… sweeping, washing utensils, cleaning the house, whatever, all those sorts, they are mothers and girls' duties'; 'most of the duties that Mummy used to do are the ones for the girls in the family… a man, our brother, cannot wash dishes or cook, he can participate in shamba work, but we sisters are doing everything that Mum was doing'.

Amisa continued that she considered that kind of a division of labour to be okay, 'it was normal'. 'Then', she pointed out. Today, she thought, 'With my education, I have awareness and I've seen that it's good to teach your children, a girl or a boy, to do all the (domestic) activities.' She also remarked that, because of all the work in the evenings, girls are not able to do any revision; therefore, they depend a lot on what they are taught in the classroom, embedding enabling and not so enabling practices and attitudes reflected in the sets of capabilities as presented in the previous chapter. It needs to be noted, however, that in Amisa's home, regardless of the clearly gendered distinction of domestic duties, in school-related matters, there was no such, 'because you're a girl or a boy' argumentation. Their parents did not even mention 'girls' and boys' schooling; rather, they talked about the schooling of their children. Besides, it was Amisa's father who enabled her to go back to school after her divorce. Amisa's father was a veterinary surgeon, working in town and staying with the family in the village only at weekends (a common lifestyle in the area where Amisa comes from, she said). While at home, he might take part in farming activities; otherwise, her mother was in charge of running the house (with the money that her father brought or sent home). When it came to financial matters, Leyla's mother was a farmer growing bananas (for food) and coffee (as a cash crop). She also used to sell some of the bananas, and their father never asked how much money she got from that; but when the coffee was ready for sale, then her father took charge. Thus, dealing with the cash crop was a clear-cut gender concern. At Leyla's home, apart from gendered farming practices, her father and brothers used to live in a separate house that 'had a floor, but my mother and we sisters slept in the mud house', she described.

Whereas Leyla and Rabia's families held clear-cut and distinctive gender ideas, in Amisa and Wema's homes, neither the family structure nor their mother's ideas promoted gender segregation. While Naomi's father read and studied a lot, and supported his daughters to do so as well, Genefa's father 'was too, just too drunk' to attend to any household duties and take responsibility for the education of the children. Obviously, the family structure influences and everyday life practices mirror the familial ideas of the roles that the girls and women are expected to fulfil, but also the value that is given to female education. Hitherto, we have listened to the research participants' narratives concerning their familial everyday life arrangements. Next, the focus is on familial attitudes – notwithstanding that they intertwine in many ways.

'I'm Giving You Education'

In discussing parental and familial perceptions of the (ir)relevance of the schooling for girls, Odaga and Heneveld (1995, 19) outlined that children's educational outcomes are a direct result of the amount of resources and the priority that parents and families attach to each child. Morley, Leach and Lugg (2009, 60–61) found that the educational success of young Ghanaians and Tanzanians was often a product of an entire family investment, comprising not only material, but also emotional, social and cultural capital. Odaga and Heneveld (ibid.) also pointed out that in the African context, educational expenses are often covered together by the parents, older siblings and other relatives. Besides, as remarked by Odaga and Heneveld (ibid.), apart from being gender differentiated, educational decision-making is often related to the birth order and number of siblings. Evidently, a complex web of networks, relationships and *ideas* affect this human capital investment behaviour. In one way or another, cost sharing, for example, was experienced by most of the women in this research, including examples of working as a house girl, schooling paid for by an older brother, support from grandparents, uncles, etc. In addition, all the participants in this research referred explicitly to birth order when I asked them to list their family members. The i) direct, and ii) indirect costs of schooling, plus iii) the insufficient number, and iv) the low quality of schools have been identified by Colclough and his colleagues (2000, 2003) as the four main factors linking the state and household level of poverty and low attendance in primary education. Yet they found that not only financial matters and the composition of families, but also a set of 'adverse cultural practices' (ibid.), which reflect gendered societal ideologies, socio-cultural ambivalence and negative perceptions, have a negative influence on parental decision making regarding the education and schooling of girls. However, there are examples of families who have actually invested in girls' education for the very same reasons that are often given as prohibitive (Odaga and Heneveld 1995, 21; see also Vavrus 2002). As one might expect from the women's narratives presented so far, and from the fact that these women had reached higher education, it is plausible to claim that there were positive ideas about the value of female education in their families. But what kind of attitudes and perceptions were attached to education and the schooling of girls and boys in the research participants' families? In other words, what kind of *an idea of Tanzanian woman* was maintained in their families, embedding the idea of education and schooling?

In the previous section, which focused on everyday life practices and (gendered) familial arrangements, we learned that the mothers of Amisa and Wema insisted that all of their children go to school. Indeed, the alternative of not being enrolled and attending classes was practically non-existent; furthermore, they expected their children not only to 'fully participate in formal education', but also to perform well and reach further, that is, to attain a particular level of education. For example, to make sure that they were actually learning, they invested a lot of money in the children's tuition. Amisa recalled being very annoyed with the extra classes after the school days, while the other children were out and playing, and not happy at all about having tuition at the library during the holidays. In their home, it was a house rule to do school assignments first and then after that, it was time for dinner and play: 'It was a deal, a rule, except on Sunday', she described. According to Amisa, their mother was very strict, but at the same time extremely supportive in school matters, insisting that 'you should go to school'. In principle, their mother had given the option not to go to school by saying that 'it's up to you if you don't want to go', but she had also explained that:

> I don't have anything to give you ... There is no inheritance that I can give you. I'm giving you education: if you want something from me, then you have to go to school.
>
> (Amisa, 36)

Thus, in theory, Amisa had the option. However, in practice, she had no choice: there was no excuse for not going to school, apart from being really sick, or some other very important reason. 'Now I understand why my mother was saying that you have to go to extra classes', Amisa admitted. She explained that because her mother was educated (Bachelor's degree), she knew the value and benefits of education, and for that reason, she also wanted to provide education for her children. 'We understood that we didn't have any alternative (togoing to school). We were very lucky'; at the end of the day, Amisa valued the actions and attitudes of their mother.

When I asked Amisa about the very best memories regarding education and schooling, they were associated with her mother, in contrast to the ones with her father. She drew a very nice picture of her sitting at the school gate and waiting for her mother to visit her 'after a month', as she had promised:

> Since the morning, I'm at the gate waiting for my mum, sitting there in the hill, and then I see my mum coming, oh, it was very nice I remember. She used to bring pilau (rice), food and small sweet things. And then she leaves you and says, 'you know, it took me a lot of time to prepare this; you have to do very well in your studies'. Yah, it was like that.
>
> (Amisa, 26)

Although their mother was very persistent in educational matters, she had once surprised Amisa, who had thought that 'okay, today I am going to die!' There was one occasion when she had been 'dodgy', as she phrased it, and not attending the 'prep', so as the sanction, she had been suspended for a while. Astonishingly, Amisa's mother 'didn't kill her', but was saying that actually the school and the teacher had been unfair to her: 'sometimes these teachers don't understand'. Hence, as long as Amisa was doing fine in school, her mother was happy. She decided that until Amisa was expelled, she would study at the library: 'Don't worry. We will go to the library together in the morning. I will pick you up during the lunch time, and then you go back. Then we'll come home...' she had instructed, and Amisa was thinking, 'Is this my real mom?'

Wema speculated that, for her mother, the idea of a Tanzanian girl and woman entailed unquestionably the idea of education. Wema's grandfather (on the mother's side) had been educated; he had been a priest, and for him 'to be educated' was such a privilege, Wema thought. Her mother and one of his uncles had a university degree, and all the other family members on the mother's side were rather highly educated and well employed. There was a big difference, according to Wema, between the families from her mother's and her father's sides. She guessed, for instance, that her stepsisters and stepbrothers (acquired after the re-marriage of her father) did not get the motivation from their father, because he and his family did not have such a high value for and experience of education. Only Wema's father had gone to school and had a degree, but his parents, for example, had never gone to school; so, Wema's stepsisters and stepbrothers ended up in the teaching profession right after primary school. In fact, their father had hoped for all of his children to be teachers, but Wema's brother wanted to be a doctor and Wema herself ended up doing her first degree in law. Consequently, he did not support their education and schooling in any way, and it was their mother, again, who paid for such things as the medical assistance college for the brother to upgrade, and to apply (successfully) for a medical degree. To sum up, the social and familial attitudes in Wema's home were remarkably positive towards the education and schooling of all of the children, girls and boys alike. In practice, Wema's mother did not have that much time to help her children with their homework – because she had to do extra work to finance the household and the education of her children – but occasionally, she was able to sit down with them, check their exercises and discuss educational issues with them.

In Leyla's opinion, the boys in the family got more support for education and schooling. They were the ones who sat down with their father in the evenings and talked: 'they (brothers) talked about the school, and other issues, what is happening in the town, and world issues, while I was with my mother in the kitchen preparing the meal', she explained rather indignantly. When Leyla had suggested to her mother that she too would like to go out and visit her friends like her older brother did, the answer had been, 'There is work to be done!' In this way, the boys were getting wider knowledge about the world and 'what is going on'; they were also given much more freedom than the girls, Leyla judged. Yet their father was very encouraging regarding education, 'different from other fathers', Leyla said, also regarding the schooling of the girls (five daughters). The reason for that was evident. Eva, Leyla's cousin, the daughter of her father's elder brother, had been educated and she had done really well. Leyla described how Eva

> managed to get a very good job; she managed to get a good husband, unlike me, (laughing); she was supporting the family... so, my father was

kind of encouraged and used to refer to her 'Look at Eva, she has done so well, I want you to be like her.'

(Leyla, 53)

Leyla explained that, in addition to the example of Eva, her parents *wanted* their children to go to school, because it was a matter of prestige for them. According to her, in the area where she comes from, a high level of education is very much valued and appreciated.[3] However, Leyla had some discussions concerning education and schooling with the teachers, but not with her parents: 'I think whatever came was okay for them', she thought. Their father might ask how they were performing, but rarely; the question of 'affording' was raised much more often. In Leyla's family, the plan was that her father would educate the elder brother and then the latter would support the rest of the (younger) siblings, and that is what they did. Thus, in Leyla's family, 'investing in education' and cost sharing was put into practice through her brother. In addition, they had learned that putting money into girls' education is not 'a lost investment'. Eva, for instance, still contributed to her biological family, although she was married to and part of another family; they had seen, broadly speaking, how those who went to school had a better standard of living and how they brought something back by supporting their families. Consequently, in Leyla's family, as many children as possible were sent to school.

Leyla's father did not 'spend all of his money for local beer!', although 'he used to drink a lot and Mother would quarrel a lot about it…', but he invested financially and morally in education. Although Tumaini was somewhat critical of her father's less active role in the family, in comparison to her mother's various efforts, she thought that her father was very positive when it came to education, and he encouraged her to study further. As explained earlier, apart from when she finished her university studies, Tumaini was born and lived in prison camps all her life. Those environments were not that encouraging regarding education and schooling, quite the contrary. Still, her father valued education and in addition, their teacher mother supported remarkably the children's opportunities to study and learn. However, they did not have much conversation concerning educational and professional options, 'like sitting down with us and telling us what to do' because, Tumaini remarked, her parents didn't actually know the alternatives that were available because of their own low educational backgrounds. Yet, they were not rigid in their education reasoning, because of admitting to not being aware of all the possibilities that were open to the children. Indeed, sometimes it was Tumaini instructing and telling them what could and should be done. As an example, when she told them that she wanted to continue for her Master's degree, her father said: 'I know you are good at, you can do it, so, pull up your socks!' Actually, when Tumaini finished her Bachelor's degree, her father did not come for the graduation but said that he would wait for the Master's graduation. However, apart from positive views and ideas about the education and schooling of their daughters, Tumaini thought that especially for her, the reason for being sent to school was, first of all, that she was doing so well in school: 'everybody was like, this lady is so intelligent; this lady can do much'. Consequently, she 'didn't give many headaches in terms of going up: I passed my primary education, secondary education, advanced secondary, then after I was selected to university [...] so, they could afford it. [...] They were supportive because I was not very much a burden to them', she judged. Hence, as with most of the women in this study, Tumaini's educational outcomes and choices are reducible to parental *ideas* concerning educational investment *inter alia*.

Rabia discussed her educational choices, or to be more accurate, how the choices were not really hers, but depended on what her parents could afford. She failed in Form IV exams and could not enter university; consequently, her parents decided that she would go to Teacher Training College, which she could access with her grades, and, more importantly, which was free. Therefore, until the critical Form IV examinations, Rabia had 'been lucky and doing well in school', because her parents could definitely not have afforded to send her to private schools, as she sighed. But, her parents were 'giving her education' by advising in school assignments: her stepmother in language, history and geography, and her father in mathematics, chemistry and biology. Rabia also recalled how their grandmother supported them by providing incentives: good performance was rewarded with tea with milk, poor results with porridge only.[4] According to Rabia, her biological mother was 'very rude', and if she saw any mistakes in the mathematics exercise book, for example, they would be punished. For that reason, the children actually preferred pretending not to study, but were busy playing; only when she went out did they go back to their school books.

There seems to be a consensus in the literature on education, gender and development that the influence of mothers especially is crucial for the educational advancement of girls. However, in this research, many of the women actually referred to their fathers. On the other hand, Naomi considered her mother to be her role model in 'doing anything and working hard'. In addition, Naomi said that she really admired her mother in being able to combine her different roles, being a mother, taking care of the household and working at the same time. For Rehema too, her mother was a role model in being an example of a woman who can manage her life and take care of herself independently, even after the divorce. Her mother kept insisting that Rehema work hard, because she saw education to be the only way for her to gain a good life (because of her disability); she demanded that Rehema study hard and supported her in doing so, because she thought that education would take her where she wants to be. Rehema explained how she had experienced what it means to be an object of decision-making regarding educational investments and prioritising, whether or not to be supported, due to the *idea* that is embedded in the physical *fact* of disability:

> You know, we choose to help other relatives... so, my aunts and uncle they preferred to help my twin sister more, to buy different things, books

[...] At home and at school... they liked her because she was faster than me, I am slow, that's why they liked her, they preferred her.

(Rehema, 30)

When we met, Rehema was proud of being able to 'pay back' – in contrast to the relatives' presumed idea of 'lost investment' – because she was employed (due to her educational advancement) and supporting her family financially. Rehema reflected that although her family valued and ranked education highly, their relatives did not: 'They live very different life and I wonder why they don't want to change?' She answered herself by stating, 'How they perceive things was different from us'.

It was argued previously, in the methodological chapter, how much of our sense of who we are arises from who we believe we are not. Therefore, in analysing the social and familial enabling environments, the third listening of the vcr-method focuses on characterising 'They' and 'how they perceive things'. In other words, the purpose of analyses is to find out how the women perceived 'They', and how their social and familial practices, and their *ideas* concerning the education and schooling of girls and women, differed from the women's own experiences and insights.

'They'

In learning about 'Us', concerning research participants' familial everyday life practices, arrangements and attitudes towards the value, relevance and benefits of education and schooling of girls and women, we came across expressions such as 'they were different' and 'he was different from other fathers'. Let us presume, accordingly, that 'they were different' and 'perceived things differently', and examine what kind of *idea of Tanzanian woman*, according to the research participants' experiences and insights, was socially surrounding them. I am focusing on three themes that were considered most typical and critical in making the difference between 'Us' and 'They'. These themes intersect, but are divided categorically into topics of marriage, patriarchy and poverty.

Mrs Somebody

Basically, all of the research participants considered that 'They' do not value education as the first thing for a girl, but a girl is prepared for marriage and family life. The women gave examples from the few particular (coastal) areas in Tanzania, where it is customary that very young girls are withdrawn from school and guided to adulthood: to learn how to act according to social and familial traditions, to get ready to be selected as someone's wife, and to take care of their husband. Consequently, there is no reason to study (also according to the girls' views), because working at home is more important than going to school; as long as you know how to read and write, what more to expect from school; besides, soon you will be pregnant. In the Tanga region, one of the coastal areas, and where Hanifa comes from, there is a 'habit that most of the girls are not educated, *of course*', as she said. This is in contrast to Amana and Leyla's experiences, who among a few other women of the study originated from the Kilimanjaro region, which in the Tanzanian context is one of the favourable areas for education, as previously remarked. This is precisely what Leyla meant by referring to 'a prestige' and Amana by talking about 'the inheritance'. Yet in the village where Amana grew up, although most of the parents sent all of their children to school, her father used to get comments and questions such as, 'She is for marriage; why do you educate your girls?'

In discussing the 'Tanga people', Hanifa coupled the issues of marriage, schoolwork, domestic duties and cultural beliefs and ideas. In her opinion, after school and returning home, the girls are not allowed to study any more because they had *already been at school*, so: 'why are you studying, come and do things, *what kind of a girl* you are and *what kind of a wife* you will be, if you don't practice these things now?' they were asked. Evidently, she thought, the girls were not encouraged to do anything else but stay at home, and there was no need or interest to continue education later either. The assumption 'They' had was that the girl would be married 'in any day', and consequently only the primary education ought to be *given* to her. Hanifa, for example, had friends who studied in the same class with her, but who ended their educational pathways at Std VII and got married (at approximately the age of 13). Hanifa also described how, while on the one hand, she herself is now seen as a role model by some of her friends who think that education has benefitted her and her family a lot, enabling her to also educate her children, on the other hand, there were also women who wondered, 'everyday she's at school, everyday!; when is she spending time with her husband?' 'They' thought that there was a danger that Hanifa's husband might go and look for other women to 'entertain' him, because she did not give all of her attention to her husband and home, but was studying at the university elsewhere.

Earlier, we learned how Wema's mother was expected to 'take care of the whole clan of people' when she came home from work. Generally, as illustrated by Amisa and Rabia, the *idea* is that when *he* comes home from work, *she* is there, if not to 'entertain', then to cook. Amisa acknowledged that Tanzanian 'city life' is different and the fact that families do vary, but she pondered that in the villages:

> Even my tribe, even our neighbours, no, they cannot understand seeing a man cooking; they wouldn't understand even if you are sick [...] even our friends and relatives would say 'why are you cooking, you have a wife?'
>
> (Amisa, 26)

Amisa and Rabia both saw a big difference between the city and village lives, and Amisa in particular was quite optimistic about the former, because 'things have started to change'. She admitted that changing attitudes regarding the *ideas* attached to gender is a process that will take years, but she had an understanding that there are a growing number of families where the boundary between gender roles is shifting little by little.

Today, according to Hanifa, the awareness regarding the value and importance of education, for both girls *and* boys, *also* in their region, is growing, and people are realising that it is nothing but education that most of the parents can actually give and leave to their children. Yet she thought that even today, there are not that many choices for girls and women, because there are so many people and so many customs and expectations guiding the rather non-existent choices. Genefa too talked a lot about '*even today* cultural biases' regarding common beliefs about girls' education. According to her, *most* people think that girls are for marriage and taking care of children. She raised the question of earning and increasing the wealth of the family through a bride-price (50–100 cows) and the issue of being married away to another family. Leyla explained, articulating similarly to Amana, Hanifa and Genefa, that *of course, even up to now*, the boys are given priority; because it is in the *culture* that the boys will inherit the name; it is the boy who will get the name of the father and the girl will get married and be called 'Mrs Somebody'.

Tumaini pointed out that whether or not you are 'Mrs Somebody', as a woman, 'They' suppose that you do all kinds of house chores, which men, married or not, do not do: 'you just don't see men sweeping, cooking, etc. Even today, there are men who still think that there are certain tasks which can only be done by women and certain that can only be done by men.' She also remarked that there are many differences between herself and the friends that she had when she lived in the prison camps: 'They' quit their educational pathways after primary education and 'They' got married. She mused that as a consequence of not being educated and staying at home as housewives, many of her friends have ended up having a much lower standard of living than she did: 'I have managed to buy a car, they don't have one; I have managed to rent a house, but you find them living in very poor houses; they can have only basic things, it's a life which is not good, it's a miserable life; I am doing a bit better, much much better.'

Neither Tumaini nor Wema recalled running across any of their childhood friends at university. Actually, the reason for Wema to leave Tanzania, move abroad, 'far away' as she phrased it, in the first place was to study further, and second, elsewhere than in Tanzania, in order to gain distance from the *idea of Tanzanian woman*, comprising the mould of marriage and role as a housewife. 'I didn't want this "normal"', she had decided. For Wema, there was a definite mismatch between the expectations 'They' had and *her idea* of what she would like to *become*: she preferred becoming a 'Beijing' instead of a life in accordance with 'Their' idea.

The Beijing

You remember Beijing conference?[5] So, the women who went to this conference, when they came back, they were like ... with this power! So, from that time, if you are a little bit educated and you seem to know what you want, what you want of your life, and not to be ruled over by a man, then they call you Beijing!

(Wema, 32)

The definition that Wema gave for 'the Beijing' implies interesting perspectives of the *idea of Tanzanian woman*, but also critical aspects of women's empowerment introduced and briefly discussed at the very beginning. As illustrated in Figure 2.1, the facets of 'rights, skills, critical thinking, self-awareness and image' are all embedded in the process towards improvement in the position of girls (UNDP 2001), resulting in girls' and women's improved 'autonomy', 'access and control', and finally the 'social position' above all else. Although encapsulated differently, these aspects are at the core of Wema's definition. She, in becoming something other than what was expected, and making her own 'normal', was acting against the *idea of Tanzanian woman*, which is still, as we have learned, strongly maintained in the research participants' rather patriarchal social and familial environments. According to Wema's experiences, 'They' had the idea that education is the end of being for a woman, or to be more accurate, the end of presumed beings and doings as a woman. Quite the opposite, her mother had been telling Wema and her sister that 'If a mother is stupid, then the children are stupid; so, don't care even if you get married, and your husband hasn't gone to school, he doesn't have education, or if he's not intellectual; as long as you are, it's okay.' On those grounds, and to Wema's surprise, even her mother, who had been so encouraging and supportive towards education and the schooling of her children, told Wema when she was thinking about the possibility of going for further studies, 'Okay, my daughter, I am so proud of what you have achieved but I think it's now time to stop; it's time to settle down and get a family.' However, perhaps because Wema's mother was now bearing the consequences of her earlier advice, or maybe it was due to Wema's education, her own ambitions and determination, or all of these three together, Wema had replied:

I will decide myself; I have lived enough of the expectations.

(Wema, 32)

It was interesting to notice that according to the research participants' experiences and insights, the higher education of girls and women is particularly problematic for family members (husbands and other relatives), but even more difficult for the in-laws. Genefa said that most of the husbands did not like their wives to go for education, particularly higher education, because once 'the women are educated, they become more challenging [...] that because of the education, the women do not respect their husbands who don't have that much education'. When we discussed possible strategies and actions to change this kind of attitude, Genefa proposed 'education'; I asked her to clarify whose education: should we send the husbands to school? Genefa considered that it is probably not the optimal strategy, because she had seen examples of educated men who still do not allow their wives to go to school: 'Even those with whom we are studying (Master's degree), they say "when you are educated, you are very challenging"; you know, they like to be there (pointing up), most powerful in the family, and if you are given that position after your education, they don't like it.' Consequently, the husbands refuse to allow their wives to go for further education. Genefa had also some fellow students, mothers, who were really frustrated, because they had heard some worrying news from their families (living far from them while studying at the university), how their husbands, for example, were not taking care of their children. Genefa's example is comparable to the concern of 'entertaining' the husband that was brought up in Hanifa's experiences.

Hanifa herself had faced difficulties with her in-laws, especially in doing her university studies. They did not like the idea that she would go even for her first degree, let alone Master's degree studies. Most likely, for them Hanifa's diploma education was enough, presuming that the husband would provide everything for the family anyhow; however, the bigger problem was that at that time they had only one child, which was not according to the idea that her sisters-in-law in particular had in mind. Consequently, Hanifa had to postpone her university studies for one year, but then she came to an agreement with her husband who *allowed* her to continue, as she phrased it. Amisa noted that there are some examples where the in-laws are totally depending on the (financial) support of their daughters-in-law, but in most cases, the relatives of the husband do not like to see the wives have the power that comes along with their high incomes. Amisa and Rabia described how the husbands may feel inferior if the wife is earning more, and how that is quite often the end of the marriage. This is an example of how the issues of marriage, patriarchy and poverty/income are interlinked, on the one hand, and how 'They' perceive it, on the other. The issues of marriage and patriarchy intersect in that sense too, as noted by Amisa, that many of 'Beijing's' gender advocates and activists are actually divorced. 'You know, that's the picture', she started to 'paint':

> In Tanzania, in patriarchal society, most of them (gender activists) have a very bad image: most of them are not married, they have no children... [...] People do not understand what they are talking about, equality, they don't know, [...] I mean, people say 'What!' [...] You will be like an enemy of society and in most cases they end up divorcing [...]
>
> (Amisa, 26)

It was interesting to learn how often the ideas of patriarchy and 'Beijing' manifested in experiences and stories related to kitchen and cooking. Amisa and Rabia were discussing their boyfriends and how they felt about cooking in front of their relatives and friends. Because of his upbringing, Amisa thought her boyfriend did not feel uncomfortable in 'welcoming and serving the guest, pouring something to drink, helping with the dishes'; it is also because he knows that otherwise he would lose her, she laughed. Quite differently, Rabia's boyfriend was all thumbs in the kitchen and with cooking. Rabia explained that he does *like* cooking, and would like to learn, but only if they are at home by themselves; once any friends or relatives visit them, and if Rabia is around, he doesn't enter the kitchen because 'then he's a man' (she laughed with Amisa). According to Rabia, 'They' do not want to sustain only the patriarchal façade in families, but more importantly, 'they don't want to change'. Rehema too talked about the change, and how she and her sister would have liked to see their younger brother participate in domestic duties. However, it was their father who would say 'no', because it's not according to the customs. Therefore, Rehema and her sister, as educated young women, had not managed to convince their father. But even so, their brother had learned at boarding school that there is not necessarily work that is for women and for men only. According to Rehema, he learned a lot about life from other people coming from different parts of Tanzania, and when he came home for holidays, he was actually helping with the household duties.

Rehema considers that *she* will be respected by people due to her education. She describes herself to be confident (compared to previous years) and her self-esteem to be really different: 'I can talk with anyone, discussing anything', and all this because education had 'broadened my mind'. How Rehema sees herself is interestingly similar to characteristically negative attributes attached to the definition of 'Beijing'. Still, Rehema thought that 'They' would respect her due to her capacities. Amisa's mother *was* respected, because 'They', the neighbours, were afraid of her, Amisa laughed. She specified that their neighbours were, indeed, choosing their words carefully and kind of making jokes about household work, for instance, but Amisa and her family did get the neighbours' message. But no matter if 'They' were thinking 'eh... this woman.!' Amisa was proud of her, and her being unquestionably a 'Beijing':

> She was liberated, I think, because of how they were raised, and because of her experience in her marriage.
>
> (Amisa 26)

Amisa emphasised strongly that how they lived does not mean that 'They' would do the same; no, she continued that it was more common in their neighbourhood to find boys playing outside, others doing their exercises and assignments, and the small girls going to fetch water, cooking, and doing other domestic things, and often, it is *poverty* that is used to argue certain kinds of beings and doings.

Poverty

As previously discussed, the research on education, gender and development suggests that, girls' and women's own expectations (what to study for) and the priority given to their future roles (as mothers and wives) have, for the most part, a negative bearing on their formal educational opportunities. Nevertheless, from the research participants' experiences and insights concerning 'They', we also find reasoning for gender-biased social and familial practices that are reducible to poverty. According to Hanifa's assessment, there are families who do value education, but who simply do not have the money, even to buy food, and consequently they cannot pay for schooling – even for their boys. Genefa and Leyla, interestingly the two oldest women participating in the study, emphasised the significance of poverty in parental decision-making. Amisa remembered that where she grew up, there were families who were well-off, others reasonably well-off, and others who could not afford to take their children to school, let alone to better schools so that they could actually learn (which again discourages parents from investing in education). She also considered that because of poverty, and all the time that the parents have to spend in securing the vital conditions for their family, they cannot monitor whether their children do, in fact, go to school and attend classes. In comparison to other single-mother households that Amisa had seen around her, she said that their family was doing rather well. She had known families who were abandoned by the fathers and living with very little income or even none; but her mother had managed to bring up and support them well alone, although they were also helped significantly by their grandparents and other relatives.

Rehema gave an example of how education is a means not only to improve one's standard of living but to have and live a *good* life. For instance, where she comes from, there is a river close by and people very much depend on the catch of the day, both for food and to gain some income. Through education, she proposed, 'They' could learn, for example, how to improve their fishing skills, and not only learn how to fish better (i.e., to get a bigger catch), but also to know how to cook in such a way that the fish retains more nutrients, and thus improve bodily health, etc. Yet, she thought that 'they don't care about education, they don't care about "life"; they just want to get food, which is not okay, because life is not all about eating and sleeping'. Rehema claimed that we all should have a good life, and education is the tool to make life better. Thus, according to Rehema, even with poverty, it is possible to improve the quality of life.

Amana was very definite in stating that because of poverty, the meaning of education is first and foremost attached to financial reasoning:

> We only value education because we can benefit from it; we get returns from investing the education to our children.
>
> (Amana, 36)

According to her experience, the girls are the ones who actually pay back what has been invested for them, and they care more about their (biological) families than the boys do. She explained how she had heard one father confessing that in their household it was the daughter who was feeding them sometimes, being able to do that exactly because of her education. Whether the parents have the awareness of the returns of educational investment or not, Leyla concluded our discussion regarding poverty by saying that the fact is that:

> When you have very little money, when it comes to choice who should be given the priority, it is always the boy.
>
> (Leyla, 53)

In my study, the interest and focus has been elsewhere than on examining the impact of poverty on gendered educational decision-making and the correlation between poverty and household arrangements, and for this reason, we have not talked a lot about money, finance and educational investments, unless brought up by the women themselves. However, because of the bearing on familial everyday life practices, it is clear that poverty and socio-economic status are *facts* that cannot be disregarded, and that *is* one of the interests and key concerns in the study, which again affects girls' and women's educational achievements and freedoms. Yet to have the freedom to achieve leads us to the questions of the utilisation of resources, opportunities and decision-making. However, prior to moving to the agency discussion, the key findings from the women's social and familial enabling environments are summarised and assessed as their social and familial sets of capabilities.

'Because I am a Girl'

The focus of this chapter has been on investigating the research participants' familial arrangements in their childhood homes, referring to, first, the everyday life practices, and second, the parental values and attitudes towards the education and schooling of their children. We heard from the women's narratives what kind of division of labour was applied in their families and maintained as per gender; we also traced the distinction that according to the women's experiences and insights separates 'Us' from 'They'. We learned that the social and familial practices and *ideas* intersect; consequently, when scratching the very uppermost surface of very practical issues such as household chores, we perceive how the *facts* are, by definition, for the most part socially constructed. Here, as in the previous chapter on the degree to which the school environment is enabling, it would be rather undemanding to conclude by listing various factors and entangling relations that constrained the research participants in functioning and pursuing their goals, but in preference, my aim is to focus on sets of opportunities that their social and familial environments opened up for them, and to realise the various educational beings and doings that the women had a reason to value. Thus, to summarise and assess women's experiences and insights, two converging questions are interrogated: what did the research participants do at home, *de facto*, before and after schooling, and how enabling were their everyday life practices and familial ideas for achieving educational well-being and agency aspirations?

One of the key assumptions in the study has been that the research partici-pants' familial arrangements, reflecting the parental ideas concerning the value and relevance of education and schooling of girls and women, were pivotal in enabling them to function, that is, to access and participate in education, and to learn and perform. What may be concluded from the previously heard narratives is that the moral encouragements, on the one hand, and financial support, on the other, that the women got from their families, were the two most critical factors enabling them to construct educational pathways, and hence, reflecting their sets of capabilities. However, regardless of the positive familial *ideas* about girls and women's education and schooling, they were not favoured in familial everyday life practices; on the contrary, and presumably the household duties diminished their opportunities and freedom for educational achievements. As was concluded in the previous chapter, which focused on the school environ-ment, the expression 'Yah, we used to work all the time', seemed to characterise the research participants' familial environments. Clearly, the answer to the question regarding what the women did at home before and after their school days was 'household chores'. They were fetching water, cooking, cleaning, sweeping, washing dishes, feeding the cattle, looking after siblings, etc. That is very much as per the *idea of Tanzanian woman*, exemplified in expressions like 'just normal stuff', 'as usual', 'every evening'. Accordingly, 'I had no time' to 'calmly sit down and study effectively in a conducive environment', 'I had too much work to do' depicts the women's familial sets of capabilities to achieve educational well-being. Yet two remarks need to be made concerning the research participants' capabilities to realise various educational functionings, the first concerns the gender aspect, and the second, the issue of adaptive preferences.

First, in some of the research participants' families, all of the family members were doing all kinds of household work, whereas in some others, there was a very clear division of labour between the sexes. As we heard from the narratives of Wema and Amisa, in their families, all the children got equal amounts of resources and opportunities to study and learn from their mothers, repre-sented by monthly allowances, and duties, responsibilities and freedoms. In contrast, according to Leyla, spending all the evenings in the kitchen con-strained her opportunities to socialise with friends and more importantly, to learn, not only about school issues but 'what is going on in the world'. Leyla was explicitly talking about freedom, referring to the opportunity to achieve the kind of knowledge that she would have valued, which in her family was given to her brother but from which she was excluded . Hanifa too considered that she and her sisters were given different kinds of opportunities to function than their brothers. She described how their mother used to tell the daughters to 'listen to your grandmother'. What the grandmother was telling them included clear guidelines of how *the girls* should behave and how they should feel, differently and in contrast to *the boys*: 'girls should be shy – that is important – and girls should not talk as loud as boys', she had instructed. Hanifa illustrated how the girls learn what they *are not* just by listening to the upbringing regarding what the boys are: as a boy, you should not cry, only girls do that; as a boy, you must be courageous, you must be stronger (than your sisters), you should not worry about anything, etc. In Hanifa's opinion, this kind of attitudinal and gendered guidance at home has a direct bearing on how girls and women later behave and feel: 'like I am not supposed to be as courageous and strong as my brother is; I shouldn't speak out and raise my voice; neither in front of other people in a classroom'. Leyla too, considered that the familial *idea* of *being* a girl, and *doing* accordingly, which meant for her i) to be in time, ii) to not to be lazy, and iii) to work hard, impacted on female students' school behaviour and was maintained in the school as well. Hence, the structures of the school system were for Leyla to sustain rather than to challenge the gendered and culturally conditioned status quo (see e.g. Morley 2010 in debating the role of education). Hanifa summarised the gendered familial arrangements and somewhat non-existent freedoms (intertwining with the school-related well-being) by arguing that a particular kind of behaviour is presumed and the other kind unforeseen:

Because I am a Girl[6] [...]. Internally, you have it in your mind.

(Hanifa, 33)

The second remark concerns the issue of adaptation, mirrored in the women's expressions such as 'just normal' and 'as usual', which may be interpreted to signify preferences that were adapted in accordance with what was seen as possible (see e.g. Unterhalter 2012). In other words, it was as per the *idea of Tanzanian woman* to give priority for the familial affairs at the cost of school-related achievements, for which reason all the work to be carried out 'every evening' was comprehended as 'just normal'. To 'work all the time' was not, however, in line with the women's agency aspirations, as we learned from their narratives concerning the boarding school environments at the time of secondary education. Thus, getting the experience of freedom to focus on studying and learning gave some perspective to the women's previous experiences regarding schooling, opportunities to function and pursuing their agency goals, which their familial everyday life practices were constraining. For this reason, expressions such as, 'during holidays, I had too much work to be done' and 'I had no time for reading' were not taken as a given or as normal as earlier, but stated in a somewhat more critical tone. As an anecdote, Amisa would have preferred to do anything else except sitg in the library and have tuition, that is, *de facto*, to have the resources and opportunities to study and learn, but as she herself admitted, the agency goals and freedoms of a child ought to be put into proportion and scaled reasonably. On the other hand, it is important to note that, although the women explained that they spent most of their time doing domestic activities, it was not considered to be all negative, quite the opposite. Consequently, although familial arrangements might have limited the women's opportunities for educational achievements and agency freedoms alike, still the hard work at home was something that they did *with* their parents and siblings as a family, and for which reason they actually had very nice memories concerning all the familial work. Evidently, the families signified for the women a pivotal source of joy and happiness, and a valuable sense of togetherness, similar to the 'forms of capital' of emotional stability and supportive relationships, as phrased by Morley, Leach and Lugg (2009), in identifying different forms of parental support for educationally successful women in Ghana and Tanzania.

All in all, the women's everyday life practices, that is, the various kinds of household duties and responsibilities, constrained their freedoms to achieve educational well-being and agency aspirations. However, as pointed out previously, women's sets of opportunities comprised two critical issues without which it is unlikely that they would have been able to function. Let us consider first the issue of familial financial support. Most of the participants in this research defined the socio-economic standing and standard of living of their families as 'okay' – compared to other children; but still, financial constraints were emphasised by all of the women. 'It was hard, I could see', Leyla said, speaking on behalf of all the women, 'visible' in the clothes that they were wearing, and exemplified in going occasionally to boarding school without any pocket money, and having only the very basics such as uniform, shoes and soap with them. These women came from families with two to seven children; in addition, some of the families financed the schooling of other relatives, too. For instance, Rabia described how there were no problems when she went to primary school, because the fees and other costs were all very small; however, when she reached secondary education, problems arose, because at that time they had a big extended family; after secondary school, when 'everybody took their children to their own houses', life started to ease again. Just as Rabia's parents educated extended family members, most of the research participants were also financially assisted by their grandparents and other relatives and in many cases by their older siblings as well. The support was much needed, notwithstanding that in most of the women's families, there were two breadwinners, both mother and father. In many families, the mothers were not employed *per se*, but they were farming and doing some petty trading, from which the families gained extra income, which was pivotal for maintaining the household and critical to supplementing the father's salary to be invested in the education and schooling of the children – as assessed by Hanifa to enable the children to be enrolled in 'very good schools'. Evidently, it was a big financial investment for the parents to send the children to school, and education was accomplished as a 'kin initiative' in most of the women's cases.

Thus, to strive for and gain enough financial resources was a pre-requisite for the research participants' educational achievements. Yet to convert the available resources into functionings was critical and dependent on and supported by social factors, identifiable in the women's cases as their fathers and mothers and their *ideas* concerning the education and schooling of their daughters. The research literature on education, gender and development argues strongly for the importance of the role of mothers regarding the educational advancement of girls. However, as previously noted, in this research many of the women actually referred to their fathers. This notion is supported by the findings of Morley and her colleagues (2009) in suggesting that paternal capital and investment can help to make a difference to their daughters' opportunities in education (see also Vavrus 2002). The role of the fathers is seen as pivotal in 'public and professional domains', that is, in the formation of academic identity, whereas the mothers' support is attached more to the domestic and affective domains (Morley, Leach and Lugg 2009). Likewise, Warrington (2013) highlights the gendered nature of parental support and encouragement, since the involvement of fathers may have the advantage of authority, education, money and time, and in contrast, the mothers may struggle frequently with the lack of resources; yet their hard work and persistence offer an inspiration and role modelling of a different kind for their daughters.[7] These findings are altogether in line with the research participants' experiences and insights, barring the fact that the mothers of Wema and Amisa held both of the enabling parental roles. Besides, we need to remember that one of the research participants, Genefa, did not receive *de facto* any kind of parental support; instead, she, together with her older sister, took the role of a parent, and created the conditions, resources and opportunities for their own educational achievements. In fact, the misbehaviour of the fathers of Wema, Amisa and Genefa blocked their capability freedoms, and as per the idea of the capabilities approach, their sets of opportunities to realise educational functionings expanded as the role of their fathers contracted. Apart from these three fathers, and the father of Naomi, who were very much involved and interested in the educational well-being of their daughters, the rest of them did not 'interfere' in the schooling of their children, as they were doing well, but still, they were there, keeping in the background with their financial support and moral back-up. All of the research participants' mothers had mainly a positive impact on their educational aspirations, and some of them influenced their daughters' educational well-being and freedoms directly and explicitly, while others took a more indirect and implicit role, as per the findings by Morley, Leach and Lugg (2009) and Warrington (2013).

Although the *facts*, such as poverty and financial resources, are not comprehended to explain the research participants' educational advancement, it is plausible to argue on the basis of the women's narratives that they did influence, and above all, intersect with their gendered familial roles and identities. However, although the women did not give explanatory primacy to the social status of their families, the issue of 'affording' came out in expressions such as 'I'm giving you education', 'all I can leave you is education, good education'. However, as pointed out previously, resources and opportunities are not synonymous with educational well-being and equality, and for this reason, the critical issue here (and from the gender perspective) is that the research participants' parents were giving and leaving the education to all of their children, the daughters and sons alike. Consequently, the *fact* of affording and the *idea* of female education – the value and relevance of it – are intertwined. In other words, having the resources and capabilities precondition educational well-being, but as per the capabilities approach, it is more important to look at the relationship between the resources people have and what they can do with them. Accordingly, the research participants' capability sets comprised both 'financial' resources and 'moral' opportunities, but their utilisation and conversion into educational achievements were enabled by their parents.

The parental and familial idea concerning the value and relevance of female education is manifested in the moral encouragement that the women enjoyed, exemplified and reflected in the discourse dividing 'Us' from 'They', and in particular in how 'They' perceived the value and relevance of female education. As we know by now, the participants in this research were good, almost exceptionally good, students, performing very well, and one of the key assumptions in the study has been that their familial everyday life arrangements and their ideas concerning the education and schooling of girls and women were somehow *more supportive* than the ones socially surrounding them. We also know by now that for the most part, the research participants' familial everyday life practices did not provide opportunities and freedoms to function and pursue agency aspirations; yet, as Naomi, for instance, explicitly defined, she had lived in 'educational surroundings', by which she meant first and foremost moral encouragement. In contrast, Tumaini talked about the somewhat apathetic educational atmosphere in the prison camps, where she was born and lived almost all of her life. According to the research participants' experiences and insights, this kind of ethos seemed to characterise 'Their' idea of the value of education in general, and the education and schooling of girls and women in particular. On the basis of the women's narratives, we may conclude that the women perceived 'They' to be different regarding the expectations of what to study for, due to the priority given to girls' future roles as mothers and wives only, in favouring males, and promoting differentials in educational opportunities and achievements. For example, Hanifa talked about 'Tanga people' for whom the wives were 'for the needs of the husband'; consequently, there is no need or interest for the 'Tanga girls' to study; Hanifa also explained how the girls are socialised for this at home, because after school, 'They' might say: 'you have achieved enough; you have already been at school'. Clearly, the set of opportunities for educational achievements open to the 'Tanga girls' was very different from that of Hanifa, due to her family's investment in and moral support of all of their children's education, 'very good education' as we recall, regardless of the biased and gendered everyday life practices that were also maintained in their household.

Generally, 'Their' *ideas* were manifested in the notions of 'Mrs Somebody' and 'Beijing', the former signifying the importance of marriage and family, seemingly inevitable for the Tanzanian girls and women (see e.g. Morley, Leach and Lugg 2009), the latter referring to difficulties to be expected once *you allow your* daughter and wife to participate in education, particularly higher education – let alone higher than he (the husband) has. As we learned from the research participants' stories, mirrored in Hanifa's grandmother's upbringing, they came from rather patriarchal social and familial environments, holding quite an unconditional *idea of Tanzanian woman*. The idea would entail that, in the first place, girls and women are for marriage, family and taking care of children, even today; and even today, as most of the research participants emphasised several times, it is inconvenient for 'Us' but even more difficult for 'They', if the girls and women pursue becoming something else. It is just against the *idea*. For example, according to the research participants' experiences and insights, to construct educational pathways means negotiations with their own families (us), with their conjugal families (they), especially in aiming to reach the university level of education (against the idea). Apparently, the appropriate level and amount of education for girls and women is presumed for the most part as per 'sufficient', seemingly the most critical issue in this. For example, in investigating the opportunity structures for widening participation in higher education in Ghana and Tanzania, Morley, Leach and Lugg (2009) found that, in sharp contrast to the feminisation of higher education in high-income countries, women's participation continues to be low in sub-Saharan Africa because of threats of social stigmatisation. Women are at risk whether they obey the complex matrix of social rules and conventions or do not comply with the traditional roles; the former risks their educational opportunities while the latter puts them at risk socially. The challenge is that both kinds of achievements, achieving a degree and being a wife and a mother, being still someone's daughter but also a daughter in-law (and having responsibilities towards both of the families), might be in accordance to the women's agency aspirations (e.g. Latvala 2006; Posti-Ahokas and Okkolin 2015). Morley and her colleagues refer to a study by Edwards (1993 in Morley, Leach and Lugg 2009), and employ the concept of 'greedy' institutions, home and university, which the mature female students are caught in between. The participants in this research had the impression that, in general, 'They' did not endorse education, and within the broader social context that kind of an attitudinal ethos was 'normal', 'habit', 'custom' and 'part of their culture'; besides, the research participants perceived that 'They do not want to change'. In contrast, for most of the research participants' families, education and schooling was prestigious and a valuable inheritance, thus embedding a high intrinsic value.

Apart from the value, the 'two greedy institutions' might be linked to mean the instrumental value of education, referring to the idea of payback. In examining highly educated women in Nairobi, Kenya, and the ways that the women combined their academic aspirations and family life, Latvala (2006) found strong and conflicting obligations and loyalties towards their biological *and* conjugal families. Similar kinds of dual responsibilities, experienced and anticipated, are found in the participant narratives of this research. The idea of payback relates interestingly to higher education, because at that age, the socially prescribed expectations for the women are categorically related to

marriage and motherhood. Hence, to construct an academic career is comprehended to be done at the cost of family life and womanhood, and if given
the opportunity to do that, allowed, as one of the research participants
phrased it, it is assumed that at some point the exercise of agency freedom is
paid back. As a matter of fact, according to the findings by Morley, Leach
and Lugg (2009), it is compulsory and inevitable, after delaying the role as
per the *idea of Tanzanian woman*. Hence, on the one hand, the payback concerns the role of the wife and mother, but on the other, the financial support
for the conjugal family as being Mrs Somebody now, *and* for the biological
family, as a return of parental and familial educational investment. Most
often, the expectations to pay back are directed to the sons, but increasingly
to the daughters, too. The research participants' experiences and insights are
suggestive that for 'They', poverty had a strong negative bearing on girls' and
women's formal educational opportunities, whereas for 'Us', the relevance
and value of education was, indeed, seen more as an investment, and the
meaning and value of education attached to the presumed later financial
benefits. In that sense, the women were not expected to become 'Mrs Nobody'
and to forget their biological family for good.

On the other hand, echoing the findings of Morley, Leach and Lugg (2009),
the relative familial material poverty of the participants in this research did
not mean aspirational poverty. Like the poor parents in the study by Morley
and her colleagues who contributed other forms of capital, the parents of the
participants in this research wanted and were able to provide 'mental capital'
for their daughters, as 'the only thing that you can have; the only thing that I
can give you', as put into words by both Amisa's mother and Hanifa's parents.
At the same time, because of poverty, and because most of the parents just
did not have the 'monetary capital', the research participants were put into a
situation where they 'took the first chance – yah, that's what we did!' in
Rabia's words. What she means by 'the first chance' is that, regardless of the
potential embedded in the school environments (adequate material resources,
excellent teachers and supportive fellow students) and the social and familial
environments (sufficient financial resources, encouraging fathers and demanding
but 'always there' mothers), all this may not be enough, but you may end up
'going with the flow' and, indeed, taking what you are *given*, that is, the 'the
first chance'. The next chapter is about the relationship *between* resources and
opportunities, choices and decision-making, and moreover, about agency
aspirations and freedoms to act accordingly. In other words, the next chapter
looks for an answer to the questions: what can be done with the available
resources and opportunities, and who decides how to construct educational
well-being and agency?

Notes

1 Leach (2000) criticises the common definition that sees women as engaged in
  'income generation' and not in 'business'. Yet, as she remarks, women can be highly

entrepreneurial in their search for economic survival and are more often than not the sole source of family income, not merely supplementing it or earning casual pocket money.

2 Chagga is the major ethnic group in the Kilimanjaro Region of Tanzania, long known for supporting girls' education (e.g. Vavrus 2002).

3 Leyla comes from the Kilimanjaro Region, which is known for its high educational values and attainment. This is commonly linked with the influence of Lutheran and Catholic missionaries who arrived in the area in the late 19th century, yet Vavrus (2002) has traced 'the idea of girls' education' much farther into 'the past' (see also Stambach 2000).

4 Interestingly, food is identified by Morley, Leach and Lugg (2009) in a study in Ghana and Tanzania, as a currency used by some mothers to comfort, support and reward the educational experiences of the early years.

5 The UN Conferences and processes for women were discussed in Chapter 2. More about UN Women (including the Beijing process) in: http://www.un.org/womenwa tch/daw/beijing/

6 Hanifa is interestingly using, although not referring to, the slogan of a campaign by the Plan International http://plan-international.org/girls/

7 Dorothy Smith has (1989, 34) emphasised in her work a distinctive standpoint for women that marks them off from the men that is an experience of work around particular individuals, particularly children; this is an experience that is grounded in biological differences, 'but through complex institutional mediations organised as caring and serving work directed toward *particular* others or groups of others'. This, according to Smith, locates women as subjects in their *local and particular actualities* (and in consequence, outside the textually mediated discourse of sociology).

Bibliography

Brown, L.M. 1997. Performing Femininities: Listening to White Working-Class Girls in Rural Maine. *Journal of Social Issues* 54, no. 4: 683–701.

Campbell, A. 1987. Self-definition by Rejection: The Case of Gang Girls. *Social Problems* 34, no. 5: 451–466.

Colclough, C., S. Al-Samarrai, P. Rose, and M. Tembon. 2003. *Achieving Schooling for All in Africa: Costs, Commitment and Gender.* Ashbourne: Ashgate.

Colclough, C., P. Rose, and M. Tembon. 2000. Gender Inequalities in Primary Schooling. The Roles of Poverty and Adverse Cultural Practice. *International Journal of Educational Development* 20: 5–27.

Fennell, S., and M. Arnot. 2009. Decentring Hegemonic Gender Theory: The Implications for Educational Research. Working Paper No. 21. London, UK: DFID.

Kleinmann, S., and M.A. Copp. 1993. *Emotions and Fieldwork. of Qualitative Research Methods Vol. 28.* Newbury Park: Sage.

Latvala. J. 2006. *Obligations, Loyalties, Conflicts. Highly Educated Women and Family Life in Nairobi, Kenya.* PhD diss., University of Tampere.

Leach, F. 2000. Gender Implications of Development Agency Policies on Education and Training. *International Journal of Educational Development* 20: 333–347.

Morley, L. 2010. Gender Mainstreaming: Myths and Measurement in Higher Education in Ghana and Tanzania. *Compare: A Journal of Comparative and International Education* 40, no. 4: 533–550.

Morley, L., F. Leach, and R. Lugg. 2009. Democratising Higher Education in Ghana and Tanzania: Opportunity Structures and Social Inequalities. *International Journal of Educational Development* 29: 56–64.

Odaga, A. and W. Heneveld. 1995. *Girls and School in Sub-Saharan Africa, From Analysis to Action*. No. 298 of *World Bank Technical Paper*. Washington D.C.: World Bank.

Posti-Ahokas, H., and M.-A. Okkolin. 2015. Enabling and Constraining Family: Young Women Building Their Educational Paths in Tanzania. *International Journal of Community, Work and Family*. http://dx.doi.org/10.1080/13668803.2015.1047737.

Smith, D.E. 1989. Sociological Theory: Methods of Writing Patriarchy. In *Feminism and Sociological Theory*, ed. R.A. Wallace, 34–64. Newbury Park, CA: Sage.

Stambach, A. 2000. *Lessons from Mount Kilimanjaro: Schooling, Community, and Gender in East Africa*. New York and London: Routledge.

UNDP. 2001. *Learning & Information Pack: Gender Analysis*. New York: UNDP.

Unterhalter, E. 2012. Inequalities, Capabilities and Poverty in four African Countries: Girls' Voice, Schooling, and Strategies for Institutional Change. *Cambridge Journal of Education* 42: 307–325.

Vavrus, F. 2002. Making Distinctions: Privatisation and the (Un)educated Girl on Mount Kilimanjaro, Tanzania. *International Journal of Educational Development* 22: 527–547.

Warrington, M. 2013. Challenging the Status Quo: The Enabling Role of Gender Sensitive Fathers, Inspirational Mothers and Surrogate Parents in Uganda. *Educational Review*, 65, no. 4: 402–415.

10 Agency Narratives and Notions

Structure – Agency in the Educational Context

In this chapter, the focus is on the women's feelings and notions of their agency, including educational opportunities and alternatives (well-being freedoms) and autonomy to make decisions and take actions accordingly (agency freedom), that *is,* in brief, the relation between structure and agency. Earlier, referring particularly to the scientific tradition emphasising a socially constructed nature of reality, the concept of educational agency was defined to involve four constitutive elements: there is an intentional and rational agent; in a particular socio-cultural environment; aiming at something; and having at least two means for pursuing the goal. Evidently, it is plausible to claim that the research participants are rational and intentional agents, whose actions are contextualised and pre-conditioned by socio-cultural structures, institutions and meanings alike. But to what extent their educational beings and doings are determined, for instance, by the system, by which I mean, in this study, the educational establishment including educational policies and the school environment, and/or the *ideas* that 'They' and 'Us' have. Furthermore, to what extent are their educational aspirations and the ways in which to construct educational pathways pre-established and 'given'?

For instance, the findings from the studies of young adults in Ghana, India and Kenya (Arnot et al. 2012a, 2012b), Mozambique and Tanzania (Helgesson 2006) and Zambia (Hansen 2005) criticise the individualised view of people's lives, women and men, which gives emphasisis to individual autonomy as the major goal of the transition from education to employment, and from childhood family to one's own family (compare the studies by Thomson, Henderson and Holland 2003 in England and Northern Ireland). Similarly, in examining young women's life- and education-strategies in Tanzania, Posti-Ahokas (2012) has shown the strong impact of social and familial relationships on individual aspirations and their realisation. For example, pursuing and studying at the upper secondary and higher education level often overlaps with working, getting married, becoming a parent and establishing a family. Consequently, the ideas that 'Us' and 'They' have may significantly govern, even lock, people, particularly women, into social and familial responsibilities and obligations, and block the utilisation of well-being freedoms and the exercise of agency freedoms (see Latvala 2006; Morley, Leach and Lugg 2009; Posti-Ahokas and Okkolin 2015). Hence, apart from the functionings and capabilities to realise 'beings and doings', the research participants' notions of their freedoms to choose and decide, take actions accordingly, and to realise their educational aspirations are required for a fuller understanding of their overall well-being and agency. By listening to the women's narratives and analysing their choices and decision-making processes, four kinds of agency freedoms were identified: 'systemic given' (given by the education system/establishment), 'educated by someone', 'own reasoning' and 'yes, but no' kinds of strategies to decide what to reach for and how, when and under what pre-conditions. Next, each of these agency notions is discussed.

Systemic Given Agency Freedoms

As discussed and concluded earlier, financial issues acquire prominence in social and familial educational decision-making, and that is what Rabia meant by 'the first chance': that regardless of the potential and opportunities embedded in the school environments and the social and familial environments, and although you are doing well at school, still you do not have alternatives *de facto* but you take what you are given, here signified by the education and schooling system in Tanzania, where Wema described she was principally 'going with the flow'.

> It was depending on the (educational) institute that our parents can afford. It was not your choice. It was about what they could afford.
>
> (Rabia, 30)

Against the assumption and expectation of Rabia's mother, who thought that Rabia was just lazy and playing too much, she did well in school and did not fail until she completed Form IV in secondary education. As a result, there was no alternative but to go to Teacher Training College (TTC), which she could enter with her grades, and, more importantly, because it was free. When Rabia completed Form VI, she decided that she would like to go and study business administration. However, she was given the option to go to a different area of study with a scholarship, and because this alternative was not as costly as her first choice, she decided to take the first affordable chance.

Whereas Rabia comprehended that the 'monetary capital' influenced her educational alternatives and choices, Wema admitted that 'honestly, I don't know'. She thought that, until her decision to leave Dar es Salaam, and to move and study abroad, she was just 'going with the flow'. Wema recounted how she had always known that she wanted to go to university, and to get her picture on the wall next to her mother's graduation photo, but she did not know what field of study she would like to go for. As explained earlier, educational advancement in Tanzania is unequivocally based on performance in (national) exams. Technically, the attained grade level dictates not only whether you are allowed to continue in the first place, but also the area of study to which you are designated. Getting first class directs the students to study law and medicine, while the second class leads to educational studies, for instance; there are also some faculties who do not accept applicants from class three at all. For this reason, Wema's alternatives and opportunities were systemic as given by the educational establishment. Indeed, she went through her whole educational pathway in Tanzania accordingly:

> In high school, I was very good, I got first class, and it is sort of normal that if you get first class, then you go and study law. Then I just filled the forms and said, okay, I choose law, because I got the first class. [...] If you got division one, and you were studying art in the secondary school, then definitely you'll do law. So I did law.
>
> (Wema, 32)

Thus, as per the education system in Tanzania, the educational pathways are constructed in accordance with the grade divisions. Evidently, Wema was selected for an area of study that is one of the most difficult to enter; she was also studying in one of the most respected faculties in the country. Still, she felt that it was not what she really wanted. She also said that if there were any other alternatives, she wasn't aware of them. Even her mother didn't advise her any more at that level, but she was 'just filling in the forms'. 'I didn't know anything', she stated, and as a result:

> Okay, here I am, with everybody, studying law. [...] But it wasn't my choice really.
>
> (Wema, 32)

Studying at the university was not difficult for a brilliant student such as Wema. Besides, it was not hard because she did not have to plan any of her studies, or to choose and make decisions, because the terms were fixed and the curriculum predetermined from above according to the syllabus and the major subject. After she graduated – and got her picture beside that of her mother – she felt happy about gaining the degree. However, the area of study and the profession of law was not what she would have chosen, 'if I would have known', as she phrased it.

Similar to Wema, Leyla had a firm understanding about the 'systemic given' opportunities and freedoms regarding education and schooling. The answer that Leyla gave to the question 'who decides' was unquestionably 'the system', that is, your pass mark plus the government. Practically speaking, it was the school and the teacher who chose and made decisions who should continue and in what area of study. 'We did not have much choice actually', she noted and illustrated:

> For example, if we were doing well, and the country needed scientists by then, we were told to go for the science classes. And if you were not doing

well in those subjects, then you went to the arts. It was the school rule and you had to go. It was normal. We used to fear our teachers. And we thought they knew better.

(Leyla, 53)

When I asked if Leyla had any kinds of conversations at home concerning educational opportunities, alternatives and choices, she denied this explicitly by saying 'no, not those times, not those days'; instead, the parents left the government to decide 'what to do with us', is how Leyla phrased the somewhat non-existent exercise of freedoms. Although she went to school almost thirty years afterLeyla did, Rehema had very similar experiences and insights:

Intelligent students were doing science and slow learners arts subject. [...] The school makes you to study [...] they will send you [...] you may find that automatically...

(Rehema, 28)

Resembling the previous experiences, Genefa described how she ended up becoming a teacher just by going with the flow: she had been considering the option of becoming a nurse, but she 'was selected' to a TTC after Form IV and 'then automatically' she became a teacher; she explained that she 'was not aware' of other opportunities and 'there wasn't any guidance'. In turn, Amana assessed that the reason why she and all of her four sisters ended up in the teaching profession is that it was 'the easiest way' and 'the cheapest way'. Another expression to illustrate the same phenomenon that you take the first chance, which is possibly the cheapest and easiest way to just go with the flow, is given by Tumaini in describing how deciding the college at which you wish to study after secondary education is not really about deciding and choosing, rather it is about fitting; that you 'decide to choose' the college that is willing to take you. Hence, it's not really about *choosing what you want*, but instead, it's about *what has chosen you*.

Clearly, the expressions like 'it wasn't my choice', 'what chooses you', 'if I would have known', 'I did not know' and 'I wasn't aware', which the women used, signal a rather limited availability of opportunities and alternatives (well-being freedoms), let alone autonomy to make decisions and take actions accordingly (agency freedoms). On the other hand, it is difficult to criticise the system for only constraining the women's agency freedoms, since only Rabia and Genefa stated explicitly how they had at first set different agency goals, while the others did not really have any particular preferences. Yet a counterargument to this is that the women's agency aspirations were adapted as per the expectations and alternatives that were known to be realistic and *de facto* open to them. In this regard, the women's agency goals and agency achievements were in accordance with the idea of the education system in Tanzania. Again, presumably it is not as per the idea of the system *inter alia* that even the hypothetical existence of alternatives is not opened to its people, let alone the potential of its utilisation.

Apart from the systemic given and apparent 'alternatives' and 'opportunities' – defined for the most part as being socially constructed – many of the women discussed single individuals who influenced their educational well-being remarkably, agency aspirations and freedoms. Next, more *socially given* educational alternatives, opportunities and decision-making processes are brought into the discussion.

Educated by Someone

Above, Wema referred explicitly to her mother, how she was not there anymore to push her and guide her at the time of her university level studies, although earlier they did not have that much conversation about educational alternatives and options; it was only after the results in the national examinations came out that the discussions started. Nonetheless, as we learned from Wema's experiences, the lack of discussions did not mean that there was a lack of educational aspiration in her family; on the contrary, with Wema's mother, it was evident that she was persistently pursuing the education of all of her children. To mention a few examples to illustrate this: we learned how Wema was temporarily enrolled in another school in her grandparents' village during her mother's holiday, and how her mother invested financially in tuition, even for private lessons after the incident in which Wema was involved (discussed in the context of harassment in the school environment chapter). We also learned how Wema, her brother and their sister somehow felt that they could not fail because their mother was supporting them alone, for which reason her big brother, for instance, did not want to maintain any kind of 'manly ego' or patriarchal façade, but helped in all the ways that he could in their household. Furthermore, they could not fail because of the fear of their father, and because the parents had a really bad relationship (particularly after the accidental death of Wema's brother). All in all, it was according to *Wema's mother's goals*, she had decided and she was determined that all of her children were going to school. That was the ethos in the house; and the children, including Wema, did not have any alternatives to negotiate with their mother, not until she started to consider the option of pursuing a Master's degree.

Another single parent who was very 'strict and persistent' in educational issues was the mother of Amisa. 'Mum, you were a dictator', Amisa described how they laughed at her nowadays. Like the mother of Wema, *Amisa's mother had decided* to invest everything she could, financially and morally, into the education and schooling of her children. Amisa's story concerning the shopping list for school utensils and her expression 'you write and she chooses' seems to describe rather well the illusion of alternatives and decision-making power given to the children in their house. Yet, although Amisa confessed that her mother could be very mean sometimes, she said how she is now starting to realise why her mother actually acted as she did. Amisa thought that there was always a reason for her mother's behaviour: 'if she said no, it was no because of this and that; and if she said no, it wasn't going to change', she recalled. Although Amisa's childhood home was packed with imperatives given by their mother, Amisa still said that she would like to raise her own children the same way as she had been raised:

> I believe her (mother)! It's not that I think she's always right, of course not, but then, she IS always right. I think most of the times when I did otherwise, I used to regret 'I wish I could have listened to my mother'. She can put herself in other's shoes.
>
> (Amisa, 26)

Amisa described how she had seen her mother struggling to be able to provide everything for her children, including education and schoolingand, and how she would like to become just like her. She said that, of course, she did not feel good *when* mistreated, but still it was the feeling of closeness that first and foremost characterised her emotions for her mother. She laughed that one day their mother was lobbying the youngest child by proposing how nice it would be to have a doctor in the house. Hence, even though they (either) did not have that many conversations at home concerning the alternatives and opportunities to construct educational pathways, appreciation for the highest level of education was implied. As did Wema, Amisa also identified explicitly her mother as *the one* who had guided her educational choices. Yet, apart from the huge impetus of the mother for Amisa's educational achievements and freedoms, her educational pathway is suggestive of the 'systemic given', because she was an excellent student 'doing well', hence to 'be selected' to the government schools and the university. Despite the opportunities and alternatives being 'given by' the system and her mother, Amisa concluded that she was 'satisfied' and 'very happy' about her educational opportunities, alternatives and choices – so far. Amisa did not talk about any particular agency goals and achievements, let alone any constraints or limitations on her achievements, regardless of the fact that she had enjoyed *de facto* limited well-being freedoms and agency freedoms. The key here is that – so far – neither the interests of Amisa and her mother, nor their expectations and the capabilities provided by the education system, had conflicted; obviously, this seemed to impact on the *feeling of agency* (Welzel and Inglehart 2010).

The previous chapters touched upon the roles of the mothers in the other research participants' educational well-being. Rehema remembered how her mother 'insisted' that she study and work hard. Tumaini talked a lot about her mother's efforts and initiatives to support financially the well-being of the family. In addition, she used to encourage the children in their schooling, reading and exercises, by bringing home extra materials from the school where she was teaching. Rabia's stepmother was also a teacher, and she helped the children particularly in their language studies. Hanifa and Naomi described the significance of their mothers in the extra income generation for the family, but also for being their role-models in combining successfully the various roles their mothers used to have (cf. Morley, Leach and Lugg 2009; Warrington 2013).

However, barring the mothers of Wema and Amisa, the fathers in the research participants' childhood families were the ones with the significant influence on the women's educational choices and the construction of their academic careers (ibid.), yet preconditioned and framed by the educational system. Concerning the financial decision-making in their childhood households in general, and deciding whether to invest in their children's education, girls and boys, in particular, the participants in this research said that they did not know who actually decided where the family income was to be invested. We have been given some hints concerning parental roles and power in decision-making, such as 'being a housewife, employed/breadwinner', 'farming for food/cash crop', 'spending all of money for beer', etc. Nonetheless, according to the women's experiences and insights, the financial decision-making in their homes was *not* that obvious (cf. Castilla 2013).[1] Indeed, many of the research participants identified their fathers as the ones who impacted on their comprehension of their educational opportunities and alternatives, and consequently, on their choices and decision-making.

Genefa narrated earlier how her father was injured in an accident, and paralysed, metaphorically. However, this impacted on the whole family on a practical level. Genefa explained that his attitudes towards education and schooling were very unsupportive, and she recalled some instances when their father prohibited the third child in the family from going to school, literally:

> I remember … she was very bright, and the teachers kept on supporting her. … Sometimes our father locked her behind the doors so that she cannot go to the school.
>
> (Genefa, 40)

Hanifa narrated how her father had maintained a role of 'commander' with her older sisters and brothers, but turned later to a 'guidance and counselling' kind of an approach with Hanifa and her younger siblings. She remembered that earlier, the older children used to have a very strict study timetable when they came back home from school, and those routines were not negotiable. Their mother tried to persuade the younger ones by saying 'please, come and study, didn't you see what your father did to your sister?'. Hanifa was referring to her twin sister and the unfortunate incident that was discussed earlier, after which she thought their father changed.

> I don't know, maybe he was getting older, so he wasn't like that any more… I wonder if he learnt a lesson that if you punish people, they cannot understand anything.
>
> (Hanifa, 33)

Consequently, Hanifa did not have to study at home, 'he decided to leave us', she said, until Std VII; that is when her father suggested that 'now it's time to read and prepare yourself for the exams.' After reaching secondary education, and during the holidays when she returned home from boarding, Hanifa used to take her notes and exercise books, sit down with her father and study with his help. Hanifa's father was a teacher and he had *an idea* that all of his children should become teachers too, especially the girls. Indeed, two of Hanifa's sisters are teachers, but her brothers disagreed and went into engineering and mechanics (after quite big fights with their father); Hanifa too, went for the teaching profession. So, 'for me, I can say it is *the influence of my father*', Hanifa replied when I raised the issue of educational opportunities, alternatives, choices and decision-making.

Naomi's parents had their background in education and, like Hanifa's father, they too wanted Naomi to become a teacher. Naomi's father had seen her teaching in their education centre and he had said that there was lot of potential in her in disseminating knowledge and educating people. When I asked if she disagreed and did not want to go into the teaching profession, Naomi answered 'no, not really', because she too could see her own strengths:

> I could see what they (parents) say is what I know, what I am able to do, therefore I had no choice.
>
> (Naomi, 32)

Although Naomi talked about her parents, she seemed to mean her father, as previously discussed in the context of the (gendered) division of labour in her home. With him, she discussed her educational opportunities and alternatives, for instance, whether to go for arts or sciences subjects in secondary school, and later, whether to pursue becoming a journalist instead of a teacher, as she had dreamed of at some point. Her father gave Naomi the 'pros and cons' and she found that it was better for her to become a teacher. When I asked if she had ever regretted becoming a teacher – in accord with the given pros and cons – she said firmly 'no'. Naomi mentioned that nowadays her father is strongly encouraging her to study further, to go for a PhD, and he would really like to see her taking the degree outside Tanzania.

Both Hanifa and Naomi's narratives are suggestive that, indeed, the fathers of the research participants were the ones with whom the *discussions* at home went on. As also mentioned previously, in another context by Leyla, although complaining that it was not she who got the opportunity to have the conversations, she was reminded that their father was different from the other fathers, in supporting strongly the education and schooling of his daughters. Thus, like the findings of Morley, Leach and Lugg (2009) and Warrington (2013), and apart from these three explicitly mentioned fathers, quite a number of them seem to have made an impact in the 'public and professional domains' and on the women's academic identities and aspirations. In contrast to the roles of the research participants' mothers, who seem to have taken more control over the women's freedoms, the fathers' role seems to have been more like opening up the set of capabilities, and giving room for the feelings of freedom and agency (Welzel and Inglehart 2010).

Her husband was another significant other in Naomi's life, who enabled her to pursue university degree(s). She explained that their marriage is based on open discussions and the exchange of ideas to settle between themselves how to run their household. She described how her husband's input in truly looking after their home together with her has been indispensable. To give an example of the impact of her husband's support, Naomi narrated how *they* had discussions regarding her educational alternatives, and how *they* had agreed that it was really an option to continue – that a university degree was reachable and doable. Then, when she was actually doing her university studies, he used to wake up in the middle of the night to calm the crying baby, saying, 'You just go and study, I'll take care of him'. Because he was employed and posted elsewhere, Genefa's husband was not able to be there beside her to help take care of the household and their children. Yet Genefa explained how they too made all of their familial decisions together, household decisions and educational choices. She emphasised proudly the broad-mindedness of her husband by making it known that he only had one degree, in comparison to the Master's that she was soon to gain.

At the beginning, Hanifa's husband had not been that open-minded and tolerant about her university studies. In fact, he had said 'no' the first time she had started to talk about the Bachelor's degree. Thus, Hanifa had to postpone and she stayed at home for two years, during which time she had their first child. Hanifa thought that the reluctance of her husband 'to allow' her, as she phrased it, to go to university was because of his parents and sisters, and consequently Hanifa *had no choice*. After the two years, Hanifa asked again 'please, just allow me, nothing bad will happen' and then he acceded. To tell the truth, Hanifa said, he regretted the past for not letting her pursue further studies, and when the time came to decide for her Master's degree studies, she continued directly after the first degree. But to construct an educational pathway for the Master's degree was not unproblematic either because of the in-laws, again, and their idea that Hanifa was using any excuse, for instance studying, for not getting pregnant and having more children. But this time, Hanifa had decided *with her husband* that it was *their life, their family*, and *their decisions*; that only their views and ideas about their family life mattered, and no one else's:

> My husband told me, you just go! So long as we know what we have planned, you can just go.
>
> (Hanifa, 33)

We did not discuss with Hanifa what had happened in the two years between and what had changed her husband's mind. As remarked by Alkire (2008), in analysing the measures of agency and criticising the presumed causal relation between the expansion of (the notion of) agency and expansion in the availability of assets and resources, the expansion of agency freedom may be related to changes in diverse factors in a myriad of ways. For example, *he* might have taken a gender-sensitivity training course and changed *his* ideas and habits to suit *her* agency aspirations; alternatively, *they*, as a pair, might have worked out for themselves that a change in their roles might bring a better balance and overall well-being for their household. Yet from Hanifa's narrative, we may conclude that the extended family can be very influential in educational decision-making, and her experience is an example *par excellence* of the multiple roles and responsibilities embedded in the women's lives and every-day life practices in the global South. Still, she was able to expand her agency freedoms and pursue her own agency goals, as a joint initiative with her husband.

From Leyla's experiences, we learned how significantly even one family member and positive role model might contribute to the opportunities of others to pursue education. In her case, the example that her cousin Eva gave made a big impact in Leyla's father coming to an understanding that it was a rather favourable idea to also send his daughters to school. On the other hand, Leyla considered that because she did not turn out like Eva, 'who was doing well in having a good job, good husband, being able to support her family financially (payback)', her parents got discouraged and disappointed. Interestingly, when we met, Leyla was 53 years old but still she seemed to compare herself to her cousin Eva and her well-being achievements. Another single relative who was brought to the table in the decision-making conversations was Tumaini's uncle. We learned earlier in examining the performance and subject choices of the women that it was her uncle who had decided what she should study. Tumaini's father was not happy at all, because he knew she was very good in mathematics, and Tumaini was not convinced, because she knew she did not do well in biology. Still, because the uncle had more influence, through the power of his position of being a university lecturer, she went according to his advice: '*My uncle decided* that I should not take mathematics', Tumaini sum-marised, and for this reason her educational pathway was pre-determined and constructed accordingly until her Master's degree studies. At that point, it was she herself who started to decide what to pursue and how: *I decided* is the kind of talk she uses when describing the decision-making processes after her first degree. It was not only Tumaini's, but until the Master's degree studies, most of the women's educational experiences and pathways were basically, manifestations of limited well-being freedoms, let alone agency freedoms. Still, their insights, their language and their current *idea* of their agency are far from limited, quite the contrary: as per Figure 2.1, of all the aspects of self-empowerment and improvement of girls' and women's overall well-being are mirrored in the third kind of expression of capability freedoms, that is, one's own reasoning to construct educational well-being and agency.

Own Reasoning

> For my father it's okay whether I get married or not. He thinks that I can do better on my own. But my mother, she wishes that I get married and have children soon.
>
> (Tumaini, 33)

According to Tumaini, her parents' opinion is important for her even today, but in the sense that you know that you are cared for, that you are still their child. What comes to her life, the choices and decision-making concerning her studies, work and the potential future family life, it is *her decision now*: not her parents', not her uncle's, but hers. We learned from Tumaini's school experiences earlier how she used to be 'the best', and how she was enrolled in co-educational schools, different from the majority of this research's participants. In adapting the second reading of the vcr-method (I, me, we, us), it was interesting to note how she talked about her educational choices, career and decision-making. For example, *she had decided* that she would look for a job before going on to further studies; a few months after her graduation she was employed, and a few years later, she was promoted. She held the post of programme manager *until she felt* that there was a need to go for further studies. When *she decided* to apply to university, she was selected, and when we met, she was about to finish her Master's degree. While studying, she had been telling herself that:

> I should now stick to my studies first. Then, once I finish, that's when I can go to other businesses (family, husband, house)... because, sometimes, *combining* these things can make the situation worse. [...] You know, at this age, people are like 'you are late', lot of pressure, but I am concentrating on studies now. After finishing, I'll think of having other, private.
>
> (Tumaini, 33)

Tumaini explicitly raised the issue that she was able to compete with boys and men. She also pointed out that being a 'young, female, single' programme manager with older male and female subordinates has not been without challenges.

Like Tumaini, *Wema had decided* in her later educational career that from now on, if *she* sees opportunities that interest her, she will go for them, and if *she* arrives at a point where she wants to stop, then she'll stop. She said that yes, familial opinions (including her mother's, not to mention that of her husband) are important for her, but at the end of the day:

> "I make my own judgement. [...] I make my own normal, with my choices."
>
> (Wema, 32)

Earlier, we heard Wema saying that she did not 'want this normal' as per the *idea of Tanzanian woman* and that she had lived enough of others' expectations. As an adult, after her familial and educational experiences, she considered that she is now much more aware of her position as a human being, and not so much fulfilling what other people are expecting. Consequently, she is the one who is examining 'what *I can* do and what *I cannot* do', and not so much what *everybody else wants* her to do. Wema gave an example from her work experience in being posted to a tiny little NGO, and how she remembered sitting in the office thinking, 'No, I deserve better; I'll just work harder.' Thus, not being satisfied with her job or her education, she said to herself that 'no, this is not enough' and decided to leave Tanzania and study further abroad.

Wema's *voice* is possibly a few degrees more empowered than Tumaini's, but Amisa's is even stronger and more emancipated. When I asked Amisa to consider the decision-making regarding her *future choices*, she was very precise and firm, saying that there is 'no one', meaning 'no one else' but herself to decide:

> I mean, for example, what I choose to study, it's me myself. [...] I mean it depends on your capacity... on your strength... and your interest!
>
> (Amisa, 26)

When I asked 'who has guided' her educational choices *so far*, she answered explicitly that it was her family background.

> Because of my mother... because of my family background, you know, because what I've seen. [...] As a woman I need to liberate myself. I need to stand on my own. And to have the life that I want to have, I need to have a good education and a good job.
>
> (Amisa, 26)

Amisa valued her mother and her familial background for the significant contribution and impact on making her the person that she is. Yet she judged that without her education, she would not be who she is now. Also, when we talked about the knowledge and skills she had acquired for managing her life, Amisa intertwined her educational and familial environments by identifying them both: 'education has made me who I am but my family background made me know the importance of education' (and to pursue education in the first place). Amisa said that she is very proud of her educational career and her grades, but she emphasised that she had worked really hard, 'I don't know if we ever slept', she laughed with Rabia. Amisa emphasised that, in general, for most Tanzanian women, as per *the idea of a Tanzanian woman*, there is no reason to work hard, because there is always someone else working hard for you. What she meant is that the majority of the women are still raised to understand that 'men are everything' and men are the ones who are supposed to provide everything for the women and for the family.

Yet, going back to Amisa's future choices and decision-making, she said firmly that she does not want to wait for someone to provide for her. She explained, for example, how she has put her future plans and everything open to her boyfriend, saying:

> If you want me, things are like this: you have to choose! He agrees, or maybe he's pretending, but then I told him like, if you think you are going to marry me and then you are going to change, you are very wrong: I'll move out. So, I mean, he's aware.
>
> (Amisa, 26)

She continued by extrapolating *her idea of Tanzanian woman* that would mean 'liberating' other women. According to Amisa, women in the Third World, including Tanzania, are in a very difficult situation, not because they are not able, but because of their socialisation (at schools and within their social and familial environments as contextualised and illustrated in the study). Amisa stated that she would like to work *for women*, because:

> My dream is that one day, the women know who they are. I want them to know that they are capable, that they can do things like men. I hope that the coming generations will understand
>
> (Amisa, 26)

Because it is the 'most empowered' within this group of women, I have deliberately excerpted several parts of Amisa's story to exemplify *her voice* and *her idea of Tanzanian woman*. She is the youngest of the participants in this research, and although it is presumable that her familial background and experiences and her mother as a role-model influenced her voice and ideas, still, age is an issue that would be interesting to extrapolate further with a larger number of women of different ages.

Despite their age, Leyla and Genefa, the oldest women in the study, decided to go for their agency aspirations, continue their educational careers, and to have the opportunity and freedom to do that. '*I said*, wow, it's never too late; *I decided* now it's time; *I decided* to join university', is how Leyla described her choices and decision-making, an opportunity that she was not given earlier. Genefa, despite her age, the lack of familial support in her childhood, and not having time to concentrate on education as an adolescent, decided to continue to university studies as an adult. For Genefa, it was 'her own ambition' that pushed her further. And now she is the one who is educating and encouraging other women to keep on trying:

> Girls can manage if they are given support... if they are given the chance. [...] If I have managed, why can't you? It is possible, you can make it!
>
> (Genefa, 40)

Yes but No Ambivalence

As suggested at the beginning of this chapter, four kinds of capability freedoms to construct educational well-being and agency were identifiable from the research participants' narratives. The strategies of 'systemic given', 'educated by someone' and 'own reasoning' were presented above. In this last section, the focus is on women's 'yes but no' kinds of experiences and insights. This ambivalence seems to intersect specifically with the profession of teaching, touched on in passing above and discussed more below. The previous section ended on the description of notions of Genefa and Leyla's agency freedoms concerning their later educational careers by suggesting that they themselves chose what to study. I also mentioned that they were 'the elderly' of this study. As a matter of fact, Genefa was referring precisely to her age in considering whether the Master's degree that she was about to obtain would bring any changes in her life, just because of her age: 'Me, at this age, I don't know if I can change', she wondered. What she meant by 'change' was her profession. 'I can still teach or I can seek another job which is related to the teaching profession', she pondered, and when I asked if she would like to look for other work, and whether she would have chosen differently in the first place, she answered:

Of course, I wouldn't have chosen the teaching profession.

(Genefa, 40)

Genefa explained that the reason for not have chosen, 'of course', to become a teacher is because of the very low salary. She emphasised that, even though the teachers might like the work per se, the earnings discouraging people from joining the teaching profession in the first place, and then from staying in the profession. Instead, they end up going elsewhere and doing something else for a living. Genefa is exemplifying and evidencing exactly the issues of teacher recruitment and retention analysed by Towse et al. (2002), alternative income generation, and the grey areas of private teaching and tuition discussed by Wedgwood (2005) and Tao (2013), for example. For Genefa, upgrading her education was a means to increase her earnings 'for a better life', as she phrased it, which is also in line with the findings by Tao (ibid.) in identifying what Tanzanian teachers value in life, and how that may bring about their school and classroom behaviour.

A similar ambivalence characterises Leyla's experiences and insights regarding her professional career as a teacher. In contrast to Genefa, Leyla said unambiguously that she would not have chosen differently. 'I like teaching', she said, yet noted, 'even though the salary is very low'. However, when Leyla narrated more, how she saw her professional capability freedoms, then the ambiguousness arose. Likewise, she also said that she would have not chosen differently, however, she then continued:

But of course I would have wanted to go to other jobs...

(Leyla, 53)

Leyla reminded me about the selection process at the schools as per the student's performance and grade level, for which reason she did not have a choice for better prospects. Consequently, as Leyla put it: 'I had to go for teaching; I had no other better alternative then.' Interestingly she concluded that:

> Nobody likes to join teaching!
>
> (Leyla, 53)

In Leyla's opinion, the thorough dislike for becoming and working as a teacher is simply due to the low payment. However, after teaching for some years, she began to like the work and that is the reason why she said in our discussions that she would not have chosen differently. She 'grew to like it' is what she actually said. As with Genefa, Leyla too was optimistic for the future and the prospects of getting a better paying job after getting her Master's degree. She hoped to get the opportunity to change her *way* of living and raise her *standard* of living, alongside an improved and upgraded level of education. This particular agency aspiration of Leyla's was in line with the 'personal functionings' (separated categorically from the professional functionings) valued by the teachers interviewed by Tao (2013).

Whereas Leyla grew to like teaching, Hanifa 'forced' herself. She said that at first, she had no interest in teaching at all, but then, as we heard from her story, 'because of my father, of course', she became a teacher. Because of having her background in education, her first degree at university, for which she got funding, was also in education. For her second degree, she decided to 'stay in the safe side' and continue in the educational sciences. Hanifa wanted to emphasise that at this point, it was no one but herself who was guiding her educational and professional choices and pathway. She pointed out that her father, for instance, was not guiding any more but he was encouraging her; likewise, she was no longer held back but supported by her husband, as we learned in the previous section. Consequently, Hanifa said that she was happy with her educational opportunities and alternatives, and she was happy with her educational choices. However, despite all of the expressions of happiness regarding her educational background and work as a teacher, the same kind of ambivalence is found in her voice as was in Genefa and Leyla's:

> You know, there are things that have forced me to choose education. [...] I am now in MA in Education, but of course, I would have preferred Master's in linguistics. But because I'm now in education, then I have to force myself to like it.
>
> (Hanifa, 33)

Rehema mentioned that her father would have hoped to see her as a lawyer, unlike her mother, who wanted Rehema to become a teacher, along with the family tradition: she was a teacher, as was her father and Rehema's grandfather, too. Rehema had refused, saying that she didn't want to become a teacher, she just didn't like it. Today, Rehema is a teacher and she likes it. An important remark that she made is, however, that she likes teaching people but she does not like the profession. Similarly, in Tao's investigation (2013), the Tanzanian teachers often articulated as one of their occupational functionings that they valued 'being able to help students learn'. However, as did Genefa and Leyla, Rehema also raised the issue of a 'low class job' and the payment. In addition, Rehema explained that, even today, it is the government who decides where the teachers are posted and randomly shifted. Evidently, that does not improve the attractiveness of the profession (cf. Towse et al. 2002).

Becoming a lawyer was not a profession that Wema found alluring at all:

> I didn't tell anyone that I didn't like to study law… And somewhere after graduating, I felt this is not exactly what I would have done if I would have known.
>
> (Wema, 32)

As we heard earlier, she was just 'going with the flow'. However, Wema described how she started to get interested in reading on her own, going to the library, borrowing books, and all in all enjoying studying properly in doing her law studies at the university. Retrospectively, she considered that studying law was not that bad after all. She also said that she is rather happy with her educational choices and career, particularly because she, different to many women in the study, stepped aside from the flow and came to know that she may, indeed, choose differently. Hence, the difference between Wema's well-being achievements (functionings) and her personally valued achievements (agency achievements) had the impact for her that she decided not to adapt her preferences as per the experienced and anticipated idea and ambivalence, but she decided to create the freedoms for her own educational well-being and agency by herself:

> I always knew and I know now that there is a way, like solution, like possibility, to do stuff in life.
>
> (Wema, 32)

No Reason to Choose Differently?

In the chapter the focus of analysis has been on the research participants' educational choices and decision-making processes concerning their *de facto* educational opportunities and alternatives, that is, women's notions of their agency freedoms and well-being freedoms respectively. To answer the question 'who decided' what kind of an educational pathway to construct and how, in accordance with the *idea of Tanzanian woman* or as per the idea of becoming something else, and to summarise the previous analyses, the voice is given to Amisa's and Tumaini's educational and professional reflections.

The title of this section implies the insight stated by Amisa, for whom there was 'no reason to choose differently'. She told how she had wanted to become either a teacher or a nurse since secondary school, and how the teaching profession was something that she had liked from the very beginning. According to Amisa, the teaching profession had been there, 'ready for her, whenever she was ready to start'; furthermore, there had been people ready and willing to fund her to go for teacher training more than anything else; besides, nobody ever mentioned even the hypothetical availability of any other alternative. Consequently, she just found herself in this position, as she phrased it. Amisa's experiences and insights touch on all of the four capability freedoms that have been discussed earlier.

To begin with, practically all of the research participants had constructed the most part of their educational pathways according to the idea of *the education system* in Tanzania, that is, as per the expectations and assumptions concerning the student's competence and qualification on the basis of their performance and grade divisions. Hence, for most of the participants in this research, as for Amisa, teacher training and the teaching profession had been there, 'waiting for them', as a pre-determined and top-down educational pathway, given to them 'when ready' by the educational establishment. Consequently, the women's educational pathways were given, but educational alternatives to be chosen from, let alone freedoms to decide and act accordingly were not. Yet, the women perceived that their systemic given capability freedoms were 'sort of normal', for which reason, implying adaptation to what was seen as *de facto* reachable and possible, there was indeed, no reason to choose differently.

On the basis of the women's experiences and insights, it is plausible to claim that their social and familial enabling environments did not provide capability freedoms either, regardless of the fact that most of the women's functionings were enabled by financial support *and* moral encouragement from home, without which it is very unlike that the women would have managed to achieve their various educational 'beings and doings'. The women's narratives are suggestive that both financial arguments *and* opportunities, alternatives and proposals provided by the family members, and by those whom their parents supposed to know better, limited the utilisation of their opportunity freedoms and exercise of agency freedoms. As has been remarked previously, the reluctance in the study to explain various social and familial everyday practices as the result of poverty does not mean that financial matters are disregarded; that cannot be done, simply, given the prevailing poverty in the country. Yet, according to the women's experiences and insights, their familial arrangements and attitudes were not reducible to poverty, barring the *fact* that 'she said that there is only this (amount of money) and you could really see that there is only this', rather than to the culturally conditioned and constructed *idea of Tanzanian woman*. Hence, the financial reasoning and argumentation was there but even more significant influence came from the ideas of female education and schooling that the family members (with fluctuating meanings) had. Consequently, the alternatives and opportunities *inter alia*, and hence, the decision-making and choosing were virtually non-existent in the women's social familial sets of capabilities, for which reason almost all of the women ended up 'taking the first chance' and 'going with the flow' as per their familial instructions and recommendations categorising their own agency aspirations and achievements.

The 'systemic given' and 'educated by someone' kinds of capability freedoms that embed rather apparent alternatives and opportunities, choices and decision-making processes, were attached particularly to the women's early educational pathways as children and adolescents, but even to construct educational well-being further seem to have been for the most part structurally given and/or according to someone else's ideas. But eventually, above all in relation to their Master's degree studies, women's 'own reasoning' and own educational aspirations began to arise, and critical and empowered voices to be heard. For example, earlier we heard from Tumaini's experiences and her voice that she is highly *aware* of her *rights* to pursue *what* she wants and *how* she thinks it is achievable. She is also very aware that she is more than *skilled* to act according to her own idea *de facto* – and not only as an ideal aspiration. Earlier, we heard her talking about her *social position*. Also, the aspect of *autonomy* is evident in her voice. Hence, even though her educational alternatives and opportunities, and the pathway were both systemic given and, literally, familially constructed until her adulthood, thereafter she has evidently taken the lead. Furthermore, she is the one who makes decisions concerning her own life, not only her educational and professional career, but the choices concerning the private sphere. In other words, she is constructing her life and overall well-being as her agency achievements and according to *her idea*.

When I discussed with Tumaini and the other women about the meaning and relevance of education, the issues of 'rights, skills, critical thinking, self-awareness, self-image, autonomy, access and control, and social position' were implied and explicitly referred to, and as may be concluded from the 'Beijing' discussion, these aspect of empowerment were significant for many of the research participants. However, despite the aspects of empowerment that were at the time of our conversations powerfully and positively embedded in the women's self-image, their emancipation is somehow overshadowed by the kind of 'yes but no' ambivalent adaptation, evident in Tumaini's quote below.

> Ah, you know, sometimes I regret why I agreed to take what he wanted me to choose... But since I decided to do what my uncle wanted, then I'm happy with that.
>
> (Tumaini, 33)

But, the key herein seem to be 'the feeling' of agency regardless of the lack of evident empiric evidence of it (cf. Welzel and Inglehart 2010). The literature on African women and African feminism (e.g. Allman, Geiger and Musisi 2002; Mikell 1997), studies by Tanzanian feminists such as Mbilinyi and Mbughuni (1991), Mbilinyi and Omari (1996), Mukaranga and Koda (1997), and Ngaiza and Koda (1991), and other scholars such as Swantz (1985), to name just a few, have aimed to reshape the image of the 'Third World Woman' (Mohanty 1988), including the *idea of Tanzanian woman* and the essentialising concept of a 'girl child' in the education policy context (e.g. Fennell and Arnot 2009), which both embed the ontological assumptions of a homogenous group of oppressed females without agency, bringing along epistemic consequences. Echoing the non-hegemonic idea of a girl child and a Third World woman, the participants in this research did not at all describe their achievements as a deprivation of agency and their sets of capabilities as a dispossession of freedoms. On the contrary, despite the negative meaning of the education system and familial ideas on their capability freedoms *par excellence*, and apart from the modification of agency aspirations and adaptation of preferences, which was manifested in the ambivalent statements and narratives concerning their educational and professional well-being, the women had positive and emancipated ideas of themselves as agents in their own lives. Thus, from the women's narratives and voices we may conclude that the point here is not that their agency goals, preferences and achievements were presumably adapted, interpreted as agency deprivation; instead, the critical issue is that, despite the systemic, and socially and familially limited opportunities and freedoms, and against the *idea of Tanzanian woman*, these ten women, as 'intentional rational agents, capable of justified evaluation and decision-making', as per the earlier discussed categorical definition of educational agency, pursued what they saw as possible and realised the nearly impossible in the Tanzanian context. For that reason, it is plausible to claim that the kind of adaptation and ambivalence heard in Tumaini's deliberation, 'since I decided to do what my uncle wanted then I'm happy with that', is more a statement from a future-oriented and very rational agent, rather than an oppressed (Third World) Tanzanian woman.

Note

1 In studying intra-household allocation of incomes in rural Ghana, Castilla (2013) found asymmetric information between spouses in reporting their (his and her) farm sales. Men, in particular, were found to give gifts to extended family members and hence, to report less familial income than the actual earnings were.

Bibliography

Alkire, S. 2008. *Concepts and Measures of Agency.* Vol. 9 of *OPHI Working Papers.* Oxford: OPHI.
Allman, J., S. Geiger, and N. Musisi, eds. 2002. *Women in African Colonial Histories.* Bloomington: Indiana University Press.

Arnot, M., F.N. Chege, and V. Wawire. 2012b. Gendered Constructions of Citizenship: Young Kenyans' Negotiations of Rights Discourses. *Comparative Education* 48, no. 1: 87–102.

Arnot, M., R. Jeffery, L. Casely-Hayford, and C. Noronha. 2012a. Schooling and Domestic Transitions: Shifting Gender Relations and Female Agency in Rural Ghana and India. *Comparative Education* 48, no. 2: 181–194.

Castilla, C. 2013. Ties That Bind: The Kin System as a Mechanism of Income-Hiding Between Spouses in Rural Ghana. WIDER Working Paper No. 2013/007. Helsinki, Finland: UNU-WIDER.

Fennell, S., and M. Arnot. 2009. Decentring Hegemonic Gender Theory: The Implications for Educational Research. Working Paper Vol. 21. London, UK: DFID.

Hansen, T.K. 2005. Getting Stuck in the Compound: Some Odds Against Social Adulthood in Lusaka, Zambia. *Africa Today* 51, no. 4: 3–16.

Helgesson, L. 2006. *Getting Ready for Life: Life Strategies of Town Youth in Mozambique and Tanzania.* PhD diss., Umeå University.

Latvala, J. 2006. *Obligations, Loyalties, Conflicts. Highly Educated Women and Family Life in Nairobi, Kenya.* PhD diss., University of Tampere.

Mbilinyi, M., and P. Mbughuni, eds. 1991. *Education in Tanzania with a Gender Perspective: Summary Report.* Dar es Salaam and Stockholm: SIDA.

Mbilinyi, D.A., and C.K. Omari. 1996. *Gender Relations and Women's Images in the Media.* Dar es Salaam, Tanzania: Dar es Salaam University Press.

Mikell, G., ed. 1997. *African Feminism: The Politics of Survival in Sub-Saharan Africa.* Philadephia: PENN.

Mohanty, C.T. 1988. Under Western Eyes: Feminist Scholarship and Colonial Discourses. *Feminist Review* 30: 61–88.

Morley, L., F. Leach, and R. Lugg. 2009. Democratising Higher Education in Ghana and Tanzania: Opportunity Structures and Social Inequalities. *International Journal of Educational Development* 29: 56–64.

Mukaranga, F., and B. Koda. 1997. *Beyond Inequalities: Women in Tanzania.* Dar es Salaam, Tanzania: TGNP.

Ngaiza, M., and B. Koda, eds. 1991. *The Unsung Heroines: Women's Life Histories from Tanzania.* Dar es Salaam, Tanzania: WRDP Publications.

Posti-Ahokas, H. 2012. Empathy-Based Stories Capturing the Voice of Female Secondary School Students in Tanzania. *International Journal for Qualitative Studies in Education* 26, no. 10: 1277–1292.

Posti-Ahokas, H., and M.-A. Okkolin. 2015. Enabling and Constraining Family: Young Women Building Their Educational Paths in Tanzania. *International Journal of Community, Work and Family.* http://dx.doi.org/10.1080/13668803.2015.1047737.

Swantz, M.-A. 1985. *Women in Development: A Creative Role Denied?: The Case of Tanzania.* London: Hurst.

Tao, Sharon. 2013. Why Are Teachers Absent? Utilising the Capability Approach and Critical Realism to Explain Teacher Performance in Tanzania. *International Journal of Educational Development* 33: 2–14.

Thomson, R., S. Henderson, and J. Holland. 2003. Making the Most of What You've Got? Resources, Values and Inequalities in Young Women's Transitions to Adulthood. *Educational Review* 55, no. 1: 33–46.

Towse, P., D. Kent, F. Osaki, and N. Kirua. 2002. Non-graduate Teacher Recruitment and Retention: Some Factors Affecting Teacher Effectiveness in Tanzania. *Teaching and Teacher Education* 18, no. 1: 637–652.

Warrington, M. 2013. Challenging the Status Quo: The Enabling Role of Gender Sensitive Fathers, Inspirational Mothers and Surrogate Parents in Uganda. *Educational Review*, 65, no. 4: 402–415.

Wedgwood, R. 2005. Post-basic Education and Poverty in Tanzania. Working Paper Series No.1. Edinburgh: Centre of African Studies.

Welzel, C., and R. Inglehart. 2010. Agency, Values, and Well-Being: A Human Development Model. *Social Indicators Research* 97: 43–63.

Part IV

Conclusion

11 Concluding Remarks and Reflections

Reflecting Women's Stories

The purpose of my study has been to identify factors that support the construction of educational pathways from the gender perspective in the Tanzanian context. This has been done by learning from ten highly educated Tanzanian women: how did they do it? How did they manage to achieve various educational beings and doings, in contrast to the majority of Tanzanian women, and what factors enabled them to pursue and realise their educational aspirations? The research aimed at reaching beyond education and schooling as such, *and* broadening the comprehension of gender equality in education only as 'access to', 'participation in', and equal number of resources in education. Hence, the focus of analysis has been on the school environment *and* social and familial environments, *and* on defining educational advancement in terms of the achievement of educational *well-being* and the exercise of *agency*. Women's educational pathways have been understood in terms of their well-being achievements, but apart from the functionings, their opportunities and freedoms to achieve the various educational beings and doings that they have reason to value have been identified.

At present, our understanding of the gendered complexities concerning girls' and women's education in the global South, Tanzania in particular, is based on i) the national and international policy data, ii) findings from the previous research on education, gender and development, and most importantly, iii) the experiences and insights of ten highly educated women, hence, providing complemented comprehension of the hegemonic narrative and mainstream discourse of the challenges and achievements embedded in female education and schooling in the global South.

What seems to characterise the research participants' *school* experiences is the axiomatic difference between their primary and secondary schools, regarding both the facilities and the teaching and learning ethos. In brief, the research participants' primary schools were not very enabling and suited to opening up sets of capabilities for achieving such educational beings and doings as they had a reason to value. Instead, both the physical and the mental environments were rather poor: their schools were quite a distance away, the water and toilet facilities were there but the quality was poor, the food was non-existent or bad, the classrooms were poorly equipped and fights over desks and chairs were normal; both female and male teachers frequently used corporal punishment as a method of learning as well as to maintain discipline, and the teachers held somewhat biased ideas towards girls' ability to study and learn. However, poor physical school environments, many deficient experiences and the low quality of education did not constrain the research participants from functioning; instead, such school environments were regarded 'as normal' and to embrace sufficient and supportive conditions enough to stay enrolled, to perform well enough, and to progress further. Besides, the research participants had some very good and encouraging teachers who supported their educational well-beings and agency aspirations in primary education remarkably, most often, however, after reaching the level of secondary education. The teachers supported significantly women's academic attainments, but in addition, they made an impact on their lives more generally speaking in developing their understandings of themselves as females, not just as female students. In addition to teachers, the research participants' fellow students had a significant role in constructing their well-beings within the school environments: first, as a child, just to have friends to play and be with; later, as an adolescent, in enabling and supporting them to pursue agency goals and in the process of identity formation.

The majority of the women got a different perspective on 'normal' upon entering secondary education, and as noted previously, comprehended in this way, all of the prohibitive and constraining factors at the primary schools, may be reversed, and regarded, indeed, as supportive factors in constructing educational well-being and agency at secondary level, this represented by 'no more caning', 'all girls', 'good teachers' and a 'friendly environment'. As previously discussed, this is how most of the women defined, and appreciated and enjoyed, the kinds of conducive and harassment-free environments in their secondary schools – as per their agency aspirations. Although the boarding secondary schools differed significantly from the research participants' primary school experiences, particularly in the sense that for the first time they were able to focus on studying, still they were not only sitting and studying in a 'conducive learning environment' but instead, their days before and after school hours were work-loaded.

'Working hard all the time' seems to characterise also the women's everyday life practices and familial arrangements, which mirror also the kind of familial values and attitudes attached to the idea of girls' and women's education and schooling. Thus, what kind of sets of opportunities did their *social and familial environments* open up for them to realise the various educational beings and doings that they had a reason to value? The financial support, on the one hand, and moral encouragement, on the other, that the women got from their families, were the two most critical factors enabling them to construct educational pathways, and hence, reflecting their sets of capabilities. However, regardless of the positive familial *ideas* towards girls' and women's education and schooling, they were not favoured in familial everyday life practices; on the contrary. All in all, as previously discussed, the various kinds of household duties and responsibilities, which the research participants were assigned to and responsible for, did, indeed, diminish their opportunities and freedoms to achieve educational well-being and agency aspirations. However, as just said, their sets of opportunities comprised two critical issues without which it is unlikely that they would have been able to function. Thus, striving for and gaining enough financial resources was a pre-requisite for the research participants' educational achievements, yet converting the available resources into functionings was critically dependent on and supported by social factors, identifiable in the women's cases as their parents and their *ideas* concerning the education and schooling of their daughters. As pointed out previously, resources and opportunities are not synonymous with educational well-being and equality, and for this reason, the critical issue here is that the research participants' parents were providing education for all of their children, for the daughters and sons alike. Consequently, the *fact* of affording and the *idea* of female education – the value and relevance of it – are intertwined. Accordingly, their capability sets comprised of both 'financial' resources and 'moral' opportunities, but their utilisation and conversion into educational achievements were enabled by their parents. Hence, different to the research participants' broader social environments, in which 'They' held quite an unconditional *idea of Tanzanian woman* entailing that, in the first place, girls and women are for marriage, family and taking care of children; in contrast, 'Us', their own parents, gave both intrinsic and instrumental value to education, hence enabling *de facto* the women's education and schooling.

The third core aspect that has been examined concerned women's agency notions. In examining the research participants' educational choices, decision-making processes, the availability of opportunities and alternatives, and then focusing analytically on answering *who decided* what to reach for and how, four kinds of 'freedoms' were identified. These include 'systemic given', 'educated by someone', 'own reasoning' and 'yes, but no' kinds of strategies to construct educational well-being and agency. What may be concluded from the previous discussion is that the participants in this research perceived of having rather apparent and limited opportunities and alternatives (well-being freedoms), let alone autonomy to make decisions and take actions accordingly (agency freedoms); rather, their educational well-being and agency aspirations and freedoms were demarcated by the system (government, schools, teachers, pass marks), and they were socially given by the family members (mothers, fathers, other relatives, in-laws, husbands) rather univocally as per the *idea of Tanzanian woman*. On the other hand, despite the limited freedoms to exercise educational agency, at the end of the day, women's educational beings and doings were no longer maintained in accordance with the *idea of Tanzanian woman*, but instead, they themselves aimed, they reasoned, and they decided to became something else, as per their own agency goals, hence having reached a position to negotiate and decide.

Finally, there were some contradictions and ambivalence evidenced in the women's judgements regarding their well-being achievements and agency achievements, but still, the 'yes but no' kind of comprehension of their own agency was emancipated and positively connoted, and may be interpreted as statements from highly rational and realistic agents who aimed at optimising the existing opportunities and alternatives.

I cannot and will not make any generalisations concerning Tanzania, let alone the rest of sub-Saharan Africa and the global South, on the basis of the women's narratives. However, the findings from the women's experiences and insights may be posited in relation to and reflected towards similar contexts substantively. Moreover, I wish to pay attention to some viewpoints concerning the relevance and value of the research, including the overall setting and approach applied in the study methodologically.

Methodological Reflections

What is of significance to acknowledge is that *all* of the discussed factors identified by Odaga and Heneveld (1995; cf. Unterhalter et al. 2014) in the review more than twenty years ago, did impact on, support or constrain women's opportunities to construct educational well-being and agency. Consequently, none of them can be justifiably disregarded when pursuing gender equality in education. However, as pointed out by Morley, Leach and Lugg (2009), although the multiple markers of successful educational identity do not act independently of each other but clearly intersect, policy discourses tend to prioritise one 'structure of inequality' at a time, that is, to tell only one kind of story. As learned from the Tanzanian education policies discussed above, the strategic priorities and practices emphasise and are directed towards inequality, conditioned and constructed categorically by the school environments, regardless of the policy-research-practice findings that all suggest that much more attention should be given to 'structures of inequality' that reach beyond the educational establishment as such. On the other hand, the school environment, as well as the social and familial environments, can be analysed and assessed as people's sets of capabilities, which again give an alternative comprehension of and perspective on development and change. However, the policy problem seems to be in the inaccuracy of defining, for example, what is meant *de facto* by gender equality in education, which in turn leads to inaccuracy in measuring and assessing what has been achieved and what remains a challenge.

Consequently, the relevance and value of this kind of study and methodology is based, first, on the holistic approach that brings together various topics, themes and factors, and second, on the alternative information base that is interested in looking at people's opportunities and freedoms to pursue various beings and doings that they have a reason to value. Thus, the significance of my study is based essentially on the uniqueness of the women's experiences and insights that is acquired through the conceptual framework and methodology employed in the study, complementing critically the knowledge base and understanding to allow, indeed, the policy-makers to better understand the social world (Giddens 1987).

'The point of doing social research', from the practical angle as per the idea of Giddens, let alone the Gadamerian idea of the fusion of horizons, would entail navigating from the practice perspective again towards the policy discussion. Clearly, these pragmatic and argumentative ideas, which I as a researcher agree with and strongly favour, require the role and responsibility of the researcher to involve participation in the policy discussion. Giddens (ibid.) claims that one of the main contributions that social research can make to the formulation of practical policy is to establish a dialogical relation between researcher, policy-makers and those whose behaviour is the subject of study. Denzin and Lincoln (2008) remind us how social (and human) sciences are normative disciplines and always embedded with issues of value, ideology, power, desire, sexism, racism, domination, repression and control. However, accordingly, as they remark, one may demand from science 'that is committed up front to issues of social justice, equity, nonviolence, peace and universal rights' (ibid.). Respectively, and apart from the practical angle, from the ethical point of view, it is only justifiable to claim that the voices, experiences and insights of those who are the 'objects' of the various development initiatives, including educational policy 'objectives', ought to be heard and incorporated into the policy formulations. Besides, as remarked by Biggeri (2007) in discussing the role of children in building the future society, akin to the women who participated in this research, they are the ones who are contributing to shaping the future conversion factors for their children, and for other Tanzanian girls and women. As we heard, for example from Wema, Hanifa, Leyla and Genefa, they are all already significant role models for many girls in their families, for their students and for the women in their communities. Thus, in re-constructing their own social environments and dispositions, they are creating spaces of capabilities for other women too.

As per the constructionist comprehension of social reality, only such institutions and distinctions as are constructed over and over again are to remain. Respectively, social reality and its various constructions are about to change if some of the constructions are no longer maintained and enhanced, if new constructions are established and strengthened, and/or if the former constructions are re-shaped or re-constructed in re-conditioned contexts. Evidently, as discussed previously, gender inequalities in education are social constructions and need not be as they are, are quite harmful and we would be better if they were done away with or at least radically transformed (Hacking 2000). It is plausible to claim that Tanzanian education policy reforms are directed at and aiming to establish re-conditioned educational contexts for achieving gender equality (in comparison to gender parity only), and on the basis of the previous discussion, it is plausible to claim that the experiences and insights of the women who have re-constructed and re-conditioned their opportunities and freedoms would provide valuable information for policy-making to re-construct educational sets of capabilities.

There has been increasing attention paid in international educational research to the complexity of educational development, calling for broadening perspectives and deepening understanding, including what is perceived as evidence of change. The trend is towards complementary views and multi-dimensional research, although recent reviews of the existing research underline that our knowledge of the complexity in educational development is still limited (e.g. Lehtomäki et al. 2014, Little and Green 2009, McGrath 2010, 2013). Consequently, the importance of qualitative micro-level and actor-centred methodology in international educational research is recognised, and my study joins this debate. To begin with, as discussed earlier, one of the research debates is about the meaning and impact of adaptation, and more specifically, it is about the adaptation of wants, hopes and aspirations of the poor. As noted by Clark (2009), this argumentation may have potentially serious consequences for development policy and practice, as it undermines the case for listening to the voices of the poor, not to mention the entire participatory poverty movement. A similar kind of a concern, arising from a different scientific tradition, has been expressed by McNay (2000; see also 2003) in criticising feminist theory for constructing discursively the essential passivity of the subject in conceptualising gender identity. Obviously, the roots of these concerns and critiques are in the Marxian idea and conception of false consciousness adapted differently, even in contradicting ways, in the contemporary development discourse by philosophers, social theorists, development economists and practitioners (Clark 2009). Echoing the selection of the voices that are to be listened to and the idea of passivity in subject formation, Mauthner and Doucet (2003) argue how the research methods that we apply are not neutral techniques but are based on theoretical, epistemological and ontological assumptions. Hence, the research encounters, as well as the methods of data analysis, for example, are sites where some voices may be enhanced and some others silenced, relying essentially on the idea of whose voice and whose subjectivity is of relevance to be heard. Are poor women to be seen as passive 'Third World women' or as agents of their own lives and development? Their demand, in turn, to substantiate the recognition of the underdeveloped implications of epistemic position for the research processes and practices resonate with the critique of 'hegemonic gender theory and its implications for educational research', presented by Fennell and Arnot (2009), who argue for taking advantage of the strategic opportunities to bring together the diverse understandings of gender that are emerging from the different trajectories taken by academic traditions of research, as presumably and arguably the lack of critical engagement with and validation of 'southern' gender theory disadvantages precisely those countries that are the target of the recently adopted SDGs, for instance.

All in all, as discussed in the introduction, the critical voices within the research on education, gender and development, as in the research of development in general, seem to focus on the information and knowledge base to measure and assess development, for which reason the different forms of action research, participatory research and research that give emphasis to peoples 'voices' have intensified their status and seem to define contemporary development research. McNay (2000), for example, claims that the denying and exclusionary logic of the formation of gender identity results in an attenuated account of agency and leaves unexplored how individuals are endowed with the capabilities for independent reflection and action, such that their responses may involve accommodation *or* adaptation *or* contestation. Hence, the problem is, as per McNay's comprehension (ibid.), that very little room is given to the possibility for creative agency. As pointed out earlier, Sen (1992, 41) gives particular weight to the people's 'opportunity of reflective choice; genuine choices with serious options', which in turn implies that within the capabilities approach, human beings are ontologically seen as capable of reflective decision-making. In line with this one core idea of the capabilities approach, and in response to the recognition of the relevance and importance of the micro-level and actor-centred research approach, this study joins in the methodological research dialogue with the research participants' narratives and their creative process of becoming, indeed, contesting the *idea of Tanzanian woman* embedded in the rather monolithic idea of 'Third World women'. Besides, as noted by Reinharz (1992, 4), with whom I agree, feminist research practices, as any other research practices, must be recognised as a plurality; rather than there being a woman's way of knowing, or a feminist way of doing research, there are 'women's ways of knowing (referring to the book by Belenky et al. 1986). Hence, to repeat myself: 'stories matter – many stories matter' (Adichie 2009).

Nebel and Herrera Rendon (2006) assert that there is an important anthropological underpinning in Sen's notion of capabilities, by which they mean ontologically relational society. In this regard, every individual's capabilities are seen to emerge from the combination and interaction of individual-level capacities and the individual's relative position *vis-à-vis* social structures that provide reasons and resources for particular behaviour. Consequently, 'every one' is comprehended as plural singular in a condition of plurality (Nebel and Herrera Rendon 2006), resonating with what Catherine MacKinnon calls the 'universal singularity' of women's experiences (McNay 2000). It is plausible to claim that despite the 'potential singularity' of our experiences, the Tanzanian women's and my experiences, our socio-cultural horizons and conditions of plurality are essentially different from each other. Hence, it is only reasonable to ask how and to what extent I, as an interpreter of their narratives, can ever really understand them.

The idea of (self-)reflexivity, embedded in the voice-centered relational method, for instance, is not a new idea for social and human scientists, and the importance of being reflexive about how we interpret our data, our role in the analytic process and the pre-conceived ideas and assumptions we bring to our analysis is well recognised. However, for Mauthner and Doucet (1998, 2002, 2003), and for me too, the vcr-method does not represent just a method of analysis, but instead implies more profound meanings and aims. According to their (ibid. 2003 422–423) understanding, the vcr-method is informed by particular ontological and epistemological assumptions (the idea of relational ontology being at the core, interdependency between human beings embedded in a complex web of social relations, importance of social context). For this reason, we need to articulate the ontological nature of subjects and subjectivities that we are using in our research and the epistemological assumptions underpinning our methods of data analysis and construction of knowledge amongst other research methods. They (ibid.) highlight the relational and constructed nature of knowledge and the interdependence of parts of the entity. Rather than seeing the researcher, the method and the data as separate and neutral components, and the researcher (mainly) decontextualising the analysis and (mechanically) sorting, organising and indexing qualitative data, they aim at moving from a conventional implicit interpretation to an explicit analytical process. Practising reflexive analysis is for them a tool to translate abstract epistemological and ontological positioning into research practices (ibid.).

The scientific presuppositions of the vcr-method are at the essence of hermeneutics. They are an integral part of women's and gender studies; feminist scholars, for example, have actively debated the research relationships. Furthermore, according to my understanding, the standpoints presented above are prerequisites for the 'positive possibilities of interpretation', which I have aimed to apply by *doing* reflexive research, by operationalising it, instead of only *being* reflexive. This is close to the demand for democratisation of research processes and practices, seen as a response to the internal and external challenges of planning, designing and data collection phases of a study, as well as for the participatory analysis of data (Byrne et al. 2009). Clearly, all the above discussions concerning the relevance and validity of the research methodology, as well as the 'reliability' of the knowledge, are all suggestive of relationality and subjectivity, to be debated further in the research arenas apart from the very essentialist idea of whose voice(s) ought to be heard.

'Now my Picture is Beside my Mother's'

To sum up, 'to have the 'opportunity to participate in arguments', (Sen 2005), requires an agenda to be debated. Hence, to conclude, I return to the women's voices and narratives, and abridge a list of the most important issues that enabled the women to construct their educational well-being and agency. As remarked earlier, it would be somewhat easy to draw an exhaustively long list of factors that constrained the research participants' functionings and pursuit of goals that they have reason to value. Instead, and in accordance with the purpose of the research, this summative discussion engenders the most important factors that *supported* the women in constructing their educational well-being and agency. There are two notions, however, that I wish to raise. The first concerns the adaptation of the conceptual frame of the capabilities approach, in which I am certain one finds inconsistencies. Regardless of the possible conceptual confusion I do think that the approach and its key concepts have excellent explanatory power in analysis in this kind of study.

The second remark concerns the process of summarising, which I find nearly overwhelmingly painful to do for such rich data, unique experiences and vast insights narrated to me by these ten unique Tanzanian women. As pointed out by Mauthner and Doucet, this is *the* critical point at which my 'insights shape what is learned from the study' (Mauthner and Doucet 2003, 419). They (ibid.) continue: 'no matter which approach to coding and analysis one chooses, the researcher's insights into the study population have profound consequences for the outcome of the study'. I think that the research participants' narratives *are* the essence of this examination, but still it is possible to do justice to their experiences and insights by summing up, because it obliges recapitulation of the critical commonalities or exceptionalities to be 'argumented'.

Reinharz (1992, 166–167) has shown how sociologists have produced case studies from, for example, groups, events, institutions and communities, which have been criticised for being unconducive to theory testing but praised for their lively description. On the other hand, she reminds us how feminists write case studies for the very same reasons that non-feminist scholars do: to illustrate the *idea*, to explain the process of development over time, to show the limits of generalisations, to explore uncharted issues by starting with the limited case and to pose provocative questions. Clearly many of these aspects are included in this study, but most importantly, I think, these ten cases as successful educational pathways and manifestations of women's well-being and agency have heuristic value due to their exceptionality in the Tanzanian context, remembering that in 2006, for instance, when I met and discussed with the first of the research participants, only 2 per cent of Tanzanians were enrolled in higher education and of these, 30 per cent of the Master's degree students were female. Thus, we may conclude and claim on the grounds of the women's narratives that the most essential factors in constructing educational well-being and agency, rather essentialist in a sense that without them, it is unlikely that they would have been able to achieve their valuable beings and doings, reflecting their school and social and familial sets of capabilities, comprised:

The Ambience of the School

Referring in particular to the i) overall ethos in the school ii) friends and iii) lack of harassment.

The women valued their 'conducive and harassment-free school environments' implying, first, the educational surroundings that enabled them to calmly sit down and study hard and effectively; second, their fellow students in assisting, helping them to study hard and catching up from being left behind. Apart from the fellow students' support for their academic agency achievements; they signified like-minded friends with whom to have girly conversations that seemed to have a great role in the process of the women's identity formation. Thirdly, the harassment-free school was related to the single-sex (secondary) schools and (nearly) abolished corporal punishment and other physical disciplinary actions (after the primary level of education).

Good Teachers

The women's narratives included examples of remarkably good teachers who made a big impact on their lives, some in primary education, but most often in secondary education; their influence on the women's academic attainment, but also on their female subjectivity, was remarkable, explicitly and implicitly.

Sufficient School Facilities

Despite the limited amenities and resources characterising the women's first school experiences, the physical school environments and learning facilities might be regarded by the parents as embracing 'sufficient conditions' to actually enrol and keep their children at school, enabling the women to achieve basic educational functionings. The difference between research participants' primary and secondary school facilities was rather extreme and clearly influenced their well-being achievements and freedoms.

Familial Financial Support

The research participants' familial socio-economic standing and standard of living was seen as moderate or 'okay', at least in comparison to other children. However, all of the women emphasised familial financial constraints. Consequently, educating the children was a big financial investment for the parents, accomplished often as a kin initiative. Besides, apart from the salary/ salaries, other small businesses, gardening, farming, etc. were carried out to raise the family income to maintain the household and to enable the children to be enrolled in school, a good school, and to learn and perform well (private tuition). Yet as an investment, the educational achievements of the women were supposed to pay back later as a return of the financial benefits (instrumental value).

Parental Moral Encouragement

Although financial constraints did not constrain research participants' educational advancement, they did impact on and intersect with their gendered identities and familial roles. However, their material poverty did not imply aspirational poverty, and it was as per their parents' *idea* that the familial resources were to be converted into educational opportunities and achievements for all their children, girls and the boys alike. In the research participants' families, education and schooling of all of their children was prestigious and a valuable 'mental' inheritance (intrinsic value), and both fathers and mothers had significant, albeit different, roles in encouraging the women in their educational achievements and to pursue their educational goals. Apart from the support for constructing academic identities, the families signified for the women a reservoir of basic capabilities of joy, happiness and the feeling of togetherness.

In addition, there were some commonalities in the research participants' personal characteristics and states of minds that enabled them to construct educational well-being and agency, defined as their:

Brilliance, Persistency and Motivation

This research participants were, in short, brilliant students who did perform well and were, as a consequence, enable to advance further. One may say that they *were enabled by* the education system *and* the familial ideas, for which reason their agency achievements and freedoms exhibited some ambivalence. However, this does not depreciate the value and relevance of the fact that they were exceptionally good students. In addition, they used to work a lot, 'all the time', 'I don't know if we ever slept', as we remember them saying, valuing high achievement and being motivated by the idea of being the best.

Clearly, all of the factors presented above enabled and enhanced the research participants' personal abilities and academic ambitions, but still the resources and opportunities to be realised as their well-being achievements and freedoms were rather limited. For instance, Genefa was obliged to rely a lot only on her personal talent, determination and motivation. Obviously, all of the factors are also intersecting. Presumably, for example, the parents would have not enrolled, and, more importantly, kept the women enrolled in the school if not sufficiently resourced as per their judgement. On the other hand, the women themselves experienced that insufficient conditions did not constrain them in functioning, e.g. attending classes and passing, for instance, albeit that the insufficiency did not advance their capabilities, let alone their agency achievements and freedoms. On the one hand, one may also ask if it is reasonable to invest and allocate educational resources in teaching and learning materials if the teachers do not have adequate pedagogical knowledge and skills to utilise them. According to the research participants' experiences and insights, there are lot of very committed and talented teachers who are really good in their profession despite the limited resources. On the other hand, the distribution of resources seem to correlate closely and positively with the primary school leaving examination; yet it is plausible to contest the assumption that the abolishment of poverty of schools, for example, would result in the achievement of gender parity, let alone gender equality. The reason for the contestation is that the women's experiences are suggestive of the significance of familial and parental support, confirming the findings of many studies on education, gender and development, as has been discussed previously. Thus, one may ask whether the children, particularly girls, are enrolled in school, and not withdrawn, *despite* good infrastructure and other material teaching and learning resources, if the parental idea does not favour and give value and relevance to the education and schooling of their daughters. We may also ask, what is the meaning of sufficient resources, good teachers and overall ambience of the schools or familial support and encouragement if the girls themselves are not motivated but give preference to other valuable beings and doings?

Emphasised by the women several times is the *fact* that the *idea of Tanzanian woman* still exists, 'even today'; and it embeds various adverse cultural biases, meanings and social institutions alike that have an impact on female education and schooling. Thus, targeting one structure of inequality at a time is not enough; in Tanzania, this has been acknowledged already soon after independence, at the time when Leyla got her green Independence shoes. These ideas are exemplified once more by Amana, Genefa and Rehema, who interestingly give emphasis explicitly to equal opportunities, as well as finally by Amana and Wema in deliberation of the value of education for their lives.

> The girls are shyer in the school than boys, in primary, secondary, even in university, to speak out, to participate. [...] There are many reasons. The first may be the self-confidence, confidence to speak out, or maybe it's the culture that you were brought up [...] if the one stick in your mind, then you'll never speak out in public and you're afraid to fail. It's the shyness that girls feel.
>
> (Amana, 36)

> More importantly parents should threat their children equally at home. In most families girls are more engaged in domestic duties than boys. After school girls are engaged in cooking, washing... while their brothers are relaxing and playing. This makes them not to have enough time for studying at home. Thus parents should give all children equal opportunities.
>
> (Genefa, 40)

> When I have a family, I want every person to have equal opportunity.
>
> (Rehema, 28)

> I find that I have peace in my life. I am happy.
>
> (Amana, 36)

> Not all of my friends have gone to the university and I would not feel like superior, but there is this thing that I can feel like pride for my mother, also like somehow give me some kind of satisfaction that okay, I have made it this far, not just for me but also for her, because she pushed me and somehow it raised my self-esteem. [...] And now my picture is besides my mother's. Yah, I'm happy.
>
> (Wema, 32)

Bibliography

Adichie, C.N. 2009. The Danger of a Single Story. TEDGlobal Talk. July 2009. Interactive transcript. https://www.ted.com/talks/chimamanda_adichie_the_danger_of_a_single_story.

Belenky, M.F., B.M. Clinchy, N.R. Goldberger, and J.M. Tarule. 1986. *Women's Ways of Knowing*. New York: Basic Books.

Biggeri, M. 2007. Children's Valued Capabilities. In *Amartya Sen's Capability Approach and Social Justice in Education*, ed. M. Walker and E. Unterhalter, 197–214. New York: Palgrave Macmillan.

Byrne, A., J. Canavan, and M. Millar. 2009. Participatory Research and the Voice-Centered Relational Method of Data Analysis: Is It Worth It? *International Journal of Social Research Methodology* 12, no. 1: 67–77.

Clark, D.A. 2009. Adaptation, Poverty and Well-Being: Some Issues and Observations with Special Reference to the Capability Approach and Development Studies. *Journal of Human Development and Capabilities* 10, no. 1: 21–42.

Denzin, N.K., and Y.S. Lincoln. 2008. *Strategies of Qualitative Inquiry*. 3rd ed. London: Sage.

Fennell, S., and M. Arnot. 2009. Decentring Hegemonic Gender Theory: The Implications for Educational Research. Working Paper No. 21. London, UK: DFID.

Giddens, A. 1987. *Social Theory and Modern Sociology*. Cambridge, UK: Polity Press.

Hacking, I. 2000. *The Social Construction of What?* Cambridge, MA: Harvard University Press.

Lehtomäki, E., H. Janhonen-Abruquah, M. Tuomi, M.-A. Okkolin, M., H. Posti-Ahokas, and P. Palojoki. 2014. Research to Engage Voices on the Ground in Educational Development. *International Journal of Educational Development* 35: 37–43.

Little, A.W., and A. Green. 2009. Successful Globalisation, Education and Sustainable Development. *International Journal of Educational Development* 29, no. 2: 166–174.

Mauthner, N., and A. Doucet. 1998. Reflections on a Voice-Centered Relational Method: Analysing Maternal and Domestic Voices. In *Feminist Dilemmas in Qualitative Research: Private Lives and Public Texts*, ed. J. Ribbens and R. Edwards, 119–146. London: Sage.

Mauthner, N., and A. Doucet. 2002. Knowing Responsibly: Linking Ethics, Research Practice and Epistemology. In *Ethics in Qualitative Research*, ed. M. Mauthner, M. Birch, J. Jessop, and T. Miller, 123–145. London: Sage.

Mauthner, N., and A. Doucet. 2003. Reflexive Accounts and Accounts of Reflexivity in Qualitative Data Analysis. *Sociology* 37, no. 3: 412–431.

McGrath, S. 2010. Education and development: thirty years of continuity and change. *International Journal of Educational Development* 30, no. 6: 537–543.

McGrath, S. 2013. The weight and breadth of evidence: Some reflections on the strengths of international education and development research. *International Journal of Educational Development* 33, no. 1: 1.

McNay, L. 2000. *Gender and Agency: Reconfiguring the Subject in Feminist and Social Theory*. Cambridge, UK: Polity Press.

McNay, L. 2003. Agency, Anticipation and Indeterminacy in Feminist Theory. *Feminist Theory* 4, no. 2: 139–148.

Morley, L., F. Leach, and R. Lugg. 2009. Democratising Higher Education in Ghana and Tanzania: Opportunity Structures and Social Inequalities. *International Journal of Educational Development* 29: 56–64.

Nebel, M., and T. Herrera Rendon. 2006. A Hermeneutic of Amartya Sen's Concept of Capability. *International Journal of Social Economics* 33, no. 10: 710–722.

Odaga, A. and W. Heneveld. 1995. Girls and School in Sub-Saharan Africa, From Analysis to Action. No. 298 of *World Bank Technical Paper.* Washington D.C.: The World Bank

Reinharz, S. 1992. *Feminist Methods in Social Research.* Oxford: Oxford University Press.

Sen, A. 1992. *Inequality Re-examined.* Oxford: Oxford University Press.

Sen, A. 2005. *The Argumentative Indian: Writings on Indian Culture, History and Identity.* London, UK: Penguin.

Unterhalter, E., A. North, M. Arnot, J. Parkes, C. Lloyd, L. Moletsane, E. Murphy-Graham, and M. Saito. 2014. *Interventions to enhance girls' education and gender equality. Education Rigorous Literature Review.* London: Department for International Development.

Appendix A1

Millennium Development Goals (MDGs)

(and selected gender **targets** and *indicators*)

Goal 1

Eradicate extreme poverty and hunger.

Goal 2

Achieve universal primary education.
 Ensure that, by 2015, children everywhere, boys and girls alike, will be able to complete a full course of primary schooling:

- *Net enrolment ratio in primary education*
- *Proportion of pupils starting grade 1 who reach last grade of primary education*
- *Literacy rate of 15–24 year-olds, women and men*

Goal 3

Promote gender equality and empower women.
 Eliminate gender disparity in primary and secondary education, preferably by 2005, and in all levels of education no later than 2015:

- *Ratios of girls to boys in primary, secondary and tertiary education*
- *Share of women in wage employment in the non-agricultural sector*
- *Proportion of seats held by women in national parliament*

Goal 4

Reduce child mortality.

Goal 5

Improve maternal health.

Goal 6

Combat HIV/AIDS, malaria and other diseases.

Goal 7

Ensure environmental sustainability.

Goal 8

Develop a global partnership for development.

Appendix A2

Sustainable Development Goals (SDGs)

(targets for Goals 4. and 5. listed after goals)

Goal 1. End poverty in all its forms everywhere.

Goal 2. End hunger, achieve food security and improved nutrition and promote sustainable agriculture.

Goal 3. Ensure healthy lives and promote well-being for all at all ages.

Goal 4. Ensure inclusive and equitable quality education and promote lifelong learning opportunities for all.

Goal 5. Achieve gender equality and empower all women and girls.

Goal 6. Ensure availability and sustainable management of water and sanitation for all.

Goal 7. Ensure access to affordable, reliable, sustainable and modern energy for all.

Goal 8. Promote sustained, inclusive and sustainable economic growth, full and productive employment and decent work for all.

Goal 9. Build resilient infrastructure, promote inclusive and sustainable industrialization and foster innovation.

Goal 10. Reduce inequality within and among countries.

Goal 11. Make cities and human settlements inclusive, safe, resilient and sustainable.

Goal 12. Ensure sustainable consumption and production patterns.

Goal 13. Take urgent action to combat climate change and its impacts.

Goal 14. Conserve and sustainably use the oceans, seas and marine resources for sustainable development.

Goal 15. Protect, restore and promote sustainable use of terrestrial ecosystems, sustainably manage forests, combat desertification, and halt and reverse land degradation and halt biodiversity loss.

Goal 16. Promote peaceful and inclusive societies for sustainable development, provide access to justice for all and build effective, accountable and inclusive institutions at all levels.

Goal 17. Strengthen the means of implementation and revitalize the Global Partnership for Sustainable Development.

SDG Goal 4.

'Ensure inclusive and equitable quality education and promote lifelong learning opportunities for all'

4.1 By 2030, ensure that all girls and boys complete free, equitable and quality primary and secondary education leading to relevant and effective learning outcomes

4.2 By 2030, ensure that all girls and boys have access to quality early childhood development, care and pre-primary education so that they are ready for primary education

4.3 By 2030, ensure equal access for all women and men to affordable and quality technical, vocational and tertiary education, including university

4.4 By 2030, substantially increase the number of youth and adults who have relevant skills, including technical and vocational skills, for employment, decent jobs and entrepreneurship

4.5 By 2030, eliminate gender disparities in education and ensure equal access to all levels of education and vocational training for the vulnerable, including persons with disabilities, indigenous peoples and children in vulnerable situations

4.6 By 2030, ensure that all youth and a substantial proportion of adults, both men and women, achieve literacy and numeracy

4.7 By 2030, ensure that all learners acquire the knowledge and skills needed to promote sustainable development, including, among others, through education for sustainable development and sustainable lifestyles, human rights, gender equality, promotion of a culture of peace and non-violence, global citizenship and appreciation of cultural diversity and of culture's contribution to sustainable development

4.a Build and upgrade education facilities that are child, disability and gender sensitive and provide safe, non-violent, inclusive and effective learning environments for all

4.b By 2020, substantially expand globally the number of scholarships available to developing countries, in particular least developed countries, small island developing States and African countries, for enrolment in higher education, including vocational training and information and communications technology, technical, engineering and scientific programmes, in developed countries and other developing countries

4.c By 2030, substantially increase the supply of qualified teachers, including through international cooperation for teacher training in developing countries, especially least developed countries and small island developing States

SDG Goal 5.

'Achieve gender equality and empower all women and girls'

5.1 End all forms of discrimination against all women and girls everywhere

5.2 Eliminate all forms of violence against all women and girls in the public and private spheres, including trafficking and sexual and other types of exploitation

5.3 Eliminate all harmful practices, such as child, early and forced marriage and female genital mutilation

5.4 Recognize and value unpaid care and domestic work through the provision of public services, infrastructure and social protection policies and the promotion of shared responsibility within the household and the family as nationally appropriate

5.5 Ensure women's full and effective participation and equal opportunities for leadership at all levels of decision-making in political, economic and public life

5.6 Ensure universal access to sexual and reproductive health and reproductive rights as agreed in accordance with the Programme of Action of the International Conference on Population and Development and the Beijing Platform for Action and the outcome documents of their review conferences

5.a Undertake reforms to give women equal rights to economic resources, as well as access to ownership and control over land and other forms of property, financial services, inheritance and natural resources, in accordance with national laws

5.b Enhance the use of enabling technology, in particular information and communications technology, to promote the empowerment of women

5.c Adopt and strengthen sound policies and enforceable legislation for the promotion of gender equality and the empowerment of all women and girls at all levels

Appendix A3

EFA Goals

Aim to meet the learning needs of all children, youth and adults by 2015.

Goal 1

Expanding and improving comprehensive early childhood care and education, especially for the most vulnerable and disadvantaged children.

Goal 2

Ensuring that by 2015 all children, particularly girls, children in difficult circumstances and those belonging to ethnic minorities, have access to, and complete, free and compulsory primary education of good quality.

Goal 3

Ensuring that the learning needs of all young people and adults are met through equitable access to appropriate learning and life-skills programmes.

Goal 4

Achieving a 50 per cent improvement in levels of adult literacy by 2015, especially for women, and equitable access to basic and continuing education for all adults.

Goal 5

Eliminating gender disparities in primary and secondary education by 2005, and achieving gender equality in education by 2015, with a focus on ensuring girls' full and equal access to and achievement in basic education of good quality.

Goal 6

Improving all aspects of the quality of education and ensuring excellence of all so that recognized and measurable learning outcomes are achieved by all, especially in literacy, numeracy and essential life skills.

Appendix A4

Education 2030
Incheon Declaration and Framework for Action

Towards inclusive and equitable quality education and lifelong learning for all

1. We, Ministers, heads and members of delegations, heads of agencies and officials of multilateral and bilateral organizations, and representatives of civil society, the teaching profession, youth and the private sector, have gathered in May 2015 at the invitation of the Director-General of UNESCO in Incheon, Republic of Korea, for the World Education Forum 2015 (WEF 2015). We thank the Government and the people of the Republic of Korea for having hosted this important event as well as UNICEF, the World Bank, UNFPA, UNDP, UN Women and UNHCR, as the co-convenors of this meeting, for their contributions. We express our sincere appreciation to UNESCO for having initiated and led the convening of this milestone event for Education 2030.

2. On this historic occasion, we reaffirm the vision of the worldwide movement for Education for All initiated in Jomtien in 1990 and reiterated in Dakar in 2000 – the most important commitment to education in recent decades and which has helped drive significant progress in education. We also reaffirm the vision and political will reflected in numerous international and regional human rights treaties that stipulate the right to education and its interrelation with other human rights. We acknowledge the efforts made; however, we recognize with great concern that we are far from having reached education for all.

3. We recall the Muscat Agreement developed through broad consultations and adopted at the Global Education for All (EFA) Meeting 2014, and which successfully informed the proposed education targets of the Open Working Group on Sustainable Development Goals (SDGs). We further recall the outcomes of the regional ministerial conferences on education post-2015 and take note of the findings of the 2015 EFA Global Monitoring Report and the Regional EFA Synthesis Reports. We recognize the important contribution of the Global Education First Initiative as well as the role of governments and regional, intergovernmental and non-governmental organizations in galvanizing political commitment for education.

4. Having taken stock of progress made towards the EFA goals since 2000 and the education-related Millennium Development Goals (MDGs) as well as the lessons learned, and having examined the remaining challenges and deliberated on the proposed Education 2030 agenda and the Framework for Action as well as on future priorities and strategies for its achievement, we adopt this Declaration.

5 Our vision is to transform lives through education, recognizing the important role of education as a main driver of development and in achieving the other proposed SDGs. We commit with a sense of urgency to a single, renewed education agenda that is holistic, ambitious and aspirational, leaving no one behind. This new vision is fully captured by the proposed SDG 4 *Ensure inclusive and equitable quality education and promote lifelong learning opportunities for all* and its corresponding targets. It is transformative and universal, attends to the 'unfinished business' of the EFA agenda and the education-related MDGs, and addresses global and national education challenges. It is inspired by a humanistic vision of education and development based on human rights and dignity; social justice; inclusion; protection; cultural, linguistic and ethnic diversity; and shared responsibility and accountability. We reaffirm that education is a public good, a fundamental human right and a basis for guaranteeing the realization of other rights. It is essential for peace, tolerance, human fulfilment and sustainable development. We recognize education as key to achieving full employment and poverty eradication. We will focus our efforts on access, equity and inclusion, quality and learning outcomes, within a lifelong learning approach.

6. Motivated by our significant achievements in expanding access to education over the last 15 years, we will ensure the provision of 12 years of free, publicly funded, equitable quality primary and secondary education, of which at least nine years are compulsory, leading to relevant learning outcomes. We also encourage the provision of at least one year of free and compulsory quality pre-primary education and that all children have access to quality early childhood development, care and education. We also commit to providing meaningful education and training opportunities for the large population of out-of-school children and adolescents, who require immediate, targeted and sustained action ensuring that all children are in school and are learning.

7. Inclusion and equity in and through education is the cornerstone of a transformative education agenda, and we therefore commit to addressing all forms of exclusion and marginalization, disparities and inequalities in access, participation and learning outcomes. No education target should be considered met unless met by all. We therefore commit to making the necessary changes in education policies and focusing our efforts on the most disadvantaged, especially those with disabilities, to ensure that no one is left behind.

8. We recognize the importance of gender equality in achieving the right to education for all. We are therefore committed to supporting gender-sensitive policies, planning and learning environments; mainstreaming gender issues in

teacher training and curricula; and eliminating gender-based discrimination and violence in schools.

9. We commit to quality education and to improving learning outcomes, which requires strengthening inputs, processes and evaluation of outcomes and mechanisms to measure progress. We will ensure that teachers and educators are empowered, adequately recruited, well-trained, professionally qualified, motivated and supported within well-resourced, efficient and effectively governed systems. Quality education fosters creativity and knowledge, and ensures the acquisition of the foundational skills of literacy and numeracy as well as analytical, problem-solving and other high-level cognitive, interpersonal and social skills. It also develops the skills, values and attitudes that enable citizens to lead healthy and fulfilled lives, make informed decisions, and respond to local and global challenges through education for sustainable development (ESD) and global citizenship education (GCED). In this regard, we strongly support the implementation of the Global Action Programme on ESD launched at the UNESCO World Conference on ESD in Aichi-Nagoya in 2014. We also stress the importance of human rights education and training in order to achieve the post-2015 sustainable development agenda.

10. We commit to promoting quality lifelong learning opportunities for all, in all settings and at all levels of education. This includes equitable and increased access to quality technical and vocational education and training and higher education and research, with due attention to quality assurance. In addition, the provision of flexible learning pathways, as well as the recognition, validation and accreditation of the knowledge, skills and competencies acquired through non-formal and informal education, is important. We further commit to ensuring that all youth and adults, especially girls and women, achieve relevant and recognized functional literacy and numeracy proficiency levels and acquire life skills, and that they are provided with adult learning, education and training opportunities. We are also committed to strengthening science, technology and innovation. Information and communication technologies (ICTs) must be harnessed to strengthen education systems, knowledge dissemination, information access, quality and effective learning, and more effective service provision.

11. Furthermore, we note with serious concern that, today, a large proportion of the world's out-of-school population lives in conflict-affected areas, and that crises, violence and attacks on education institutions, natural disasters and pandemics continue to disrupt education and development globally. We commit to developing more inclusive, responsive and resilient education systems to meet the needs of children, youth and adults in these contexts, including internally displaced persons and refugees. We highlight the need for education to be delivered in safe, supportive and secure learning environments free from violence. We recommend a sufficient crisis response, from emergency response through to recovery and rebuilding; better coordinated national, regional and global responses; and capacity development for

comprehensive risk reduction and mitigation to ensure that education is maintained during situations of conflict, emergency, post-conflict and early recovery.

Implementing our Common Agenda

12. We reaffirm that the fundamental responsibility for successfully implementing this agenda lies with governments. We are determined to establish legal and policy frameworks that promote accountability and transparency as well as participatory governance and coordinated partnerships at all levels and across sectors, and to uphold the right to participation of all stakeholders.

13. We call for strong global and regional collaboration, cooperation, coordination and monitoring of the implementation of the education agenda based on data collection, analysis and reporting at the country level, within the framework of regional entities, mechanisms and strategies.

14. We recognize that the success of the Education 2030 agenda requires sound policies and planning as well as efficient implementation arrangements. It is also clear that the aspirations encompassed in the proposed SDG 4 cannot be realized without a significant and well-targeted increase in financing, particularly in those countries furthest from achieving quality education for all at all levels. We therefore are determined to increase public spending on education in accordance with country context, and urge adherence to the international and regional benchmarks of allocating efficiently at least 4–6% of Gross Domestic Product and/or at least 15–20% of total public expenditure to education.

15. Noting the importance of development cooperation in complementing investments by governments, we call upon developed countries, traditional and emerging donors, middle income countries and international financing mechanisms to increase funding to education and to support the implementation of the agenda according to countries' needs and priorities. We recognize that the fulfilment of all commitments related to official development assistance (ODA) is crucial, including the commitments by many developed countries to achieve the target of 0.7 per cent of gross national product (GNP) for ODA to developing countries. In accordance with their commitments, we urge those developed countries that have not yet done so to make additional concrete efforts towards the target of 0.7 per cent of GNP for ODA to developing countries. We also commit to increase our support to the least developed countries. We further recognize the importance of unlocking all potential resources to support the right to education. We recommend improving aid effectiveness through better coordination and harmonization, and prioritizing financing and aid to neglected sub-sectors and low income countries. We also recommend significantly increasing support for education in humanitarian and protracted crises. We welcome the Oslo Summit on Education for Development (July

2015) and call on the Financing for Development Conference in Addis Ababa to support the proposed SDG 4.

16. We call on the WEF 2015 co-convenors, and in particular UNESCO, as well as on all partners, to individually and collectively support countries in implementing the Education 2030 agenda, by providing technical advice, national capacity development and financial support based on their respective mandates and comparative advantages, and building on complementarity. To this end, we entrust UNESCO, in consultation with Member States, the WEF 2015 co-convenors and other partners, to develop an appropriate global coordination mechanism. Recognizing the Global Partnership for Education as a multi-stakeholder financing platform for education to support the implementation of the agenda according to the needs and priorities of countries, we recommend that it be part of this future global coordination mechanism.

17. We further entrust UNESCO, as the United Nations' specialized agency for education, to continue its mandated role to lead and coordinate the Education 2030 agenda, in particular by: undertaking advocacy to sustain political commitment; facilitating policy dialogue, knowledge sharing and standard setting; monitoring progress towards the education targets; convening global, regional and national stakeholders to guide the implementation of the agenda; and functioning as a focal point for education within the overall SDG coordination architecture.

18. We resolve to develop comprehensive national monitoring and evaluation systems in order to generate sound evidence for policy formulation and the management of education systems as well as to ensure accountability. We further request the WEF 2015 co-convenors and partners to support capacity development in data collection, analysis and reporting at the country level. Countries should seek to improve the quality, levels of disaggregation and timeliness of reporting to the UNESCO Institute for Statistics. We also request that the Education for All Global Monitoring Report be continued as an independent Global Education Monitoring Report (GEMR), hosted and published by UNESCO, as the mechanism for monitoring and reporting on the proposed SDG 4 and on education in the other proposed SDGs, within the mechanism to be established to monitor and review the implementation of the proposed SDGs.

19. We have discussed and agreed upon the essential elements of the Education 2030 Framework for Action. Taking into account the United Nations summit for the adoption of the post-2015 development agenda (New York, September 2015) and the outcomes of the Third International Conference on Financing for Development (Addis Ababa, July 2015), a final version will be presented for adoption and launched at a special high-level meeting to be organized alongside the 38th session of the General Conference of UNESCO in November 2015. We are fully committed to its implementation after its adoption, to inspire and guide countries and partners to ensure that our agenda is achieved.

20. Building on the legacy of Jomtien and Dakar, this Incheon Declaration is an historic commitment by all of us to transform lives through a new vision for education, with bold and innovative actions, to reach our ambitious goal by 2030.

Incheon, Republic of Korea
21 May 2015

Appendix A5

Table A5.1 Some Characteristics of Research Participants at the Time of the Interviews.

	Name	Age	Educational and Professional Background	Marital and Parental Status; Lives with
1	Amana	36	Finishing Master's degree in Education; Teacher	Divorced; children live with the father
2	Amisa	26	Finishing Master's degree in Development Studies	Cohabits with boyfriend
3	Genefa	40	Finishing Master's degree in Education; Teacher	Married with children
4	Hanifa	33	Finishing Master's degree in Education; Teacher	Married with children
5	Leyla	53	Finishing Master's degree in Education; Teacher	Divorced; dependent children
6	Naomi	32	Finishing Master's degree in Education; Teacher	Married with children
7	Rabia	30	BA in Mass Communication; University Lecturer	Cohabits with boyfriend
8	Rehema	28	B.Sc. in Home Economics and Human Nutrition; Teacher	Lives with parents
9	Tumaini	33	Finishing Master's degree in Public Health; Programme Manager (HIV/Aids)	Boyfriend; owns a house, lives alone
10	Wema	32	Finishing Master's degree in Social Sciences; Lawyer	Married

Appendix A6

Interview Themes

A) Social and Familial Environment

Living conditions (rural/urban; housing; standard of living)
Educational and professional background of the (extended) family
Family structures (gender roles)
Costs of schooling (direct and indirect costs; opportunity costs)
Perceptions about the importance of schooling
Decision-making concerning education and schooling

B) School Environment

Memories (first; best; worst)
Typical day
School infrastructure
Safety issues
Learning facilities
Teaching and learning (teachers)
Human relations (peer group)
Attainment: attendance, transitions; performance; other learning outcomes

C) Individual Motives and Decision-Making

Educational opportunities and choices
Expectations and motivation
Self-image and pushing ahead
The value and meaning of education
Future plans

D) Educational Development in Tanzania

Educational choices and experiences for own children
Critical issues from girls' perspective in general
Actions to support education and schooling of girls and women

Index

Printed and bound by CPI Group (UK) Ltd, Croydon, CR0 4YY

22/10/2024

01777623-0013